# Soybean Science and Sustainability - Productivity, Climate Resilience, and Smart Agriculture

*Edited by Ganesh Chandrakant Nikalje*

Published in London, United Kingdom

Soybean Science and Sustainability - Productivity, Climate Resilience, and Smart Agriculture
http://dx.doi.org/10.5772/intechopen.1008081
Edited by Ganesh Chandrakant Nikalje

Contributors
Alfadhl Y. Alkhaled, Anisa Nurina Aulia, Anurag Satpathi, Arini Putri Hanifa, Arlyna Budi Pustika, Atman, Deepak B. Shelke, Dhanpreet Kaur, Evert Y. Hosang, Forita Dyah Arianti, Ganesh Chandrakant Nikalje, Heni Purwaningsih, I. Gusti Komang Dana Arsana, Insan Wijaya, Kristamtini, M. Basir Nappu, Mahadev R. Chambhare, Muhammad Iskandar Ishaq, Muhammad Talha Ramzan, Niharika Kodirekka, Parul Setiya, Santosh Korav, Setyorini Widyayanti, Sugeng Widodo, Susilawati, Sutardi, Syamsul Hadi, Tayyiba Usman, Thangavel Palanivel Swarnam, Umesh Kumar Singh, Vinod Kumar, Yunish Khan, Yustisia



First published in London, United Kingdom, 2026 by IntechOpen
IntechOpen is the global imprint of INTECHOPEN LIMITED, registered in England and Wales, registration number: 11086078, 167-169 Great Portland Street, London, W1W 5PF, United Kingdom

For EU product safety concerns: IN TECH d.o.o., Prolaz Marije Krucifikse Kozulić 3, 51000 Rijeka, Croatia, info@intechopen.com or visit our website at intechopen.com.

British Library Cataloguing-in-Publication Data
A catalogue record for this book is available from the British Library

Soybean Science and Sustainability - Productivity, Climate Resilience, and Smart Agriculture
Edited by Ganesh Chandrakant Nikalje
p. cm.

This title is part of the Agricultural Sciences Book Series, Volume 37
Topic: Agronomy and Horticulture
Series Editor: W. James Grichar
Topic Editors: Murat Helvaci and Olga Panfilova

Print ISBN 978-1-83635-433-8
Online ISBN 978-1-83635-432-1
eBook (PDF) ISBN 978-1-83635-434-5
ISSN 3029-052X

If disposing of this product, please recycle the paper responsibly.

# IntechOpen Book Series
# Agricultural Sciences
## Volume 37

## Aims and Scope of the Series

The importance of agriculture cannot be overstated. It helps sustain life, as it gives us the food we need to survive and provides opportunities for economic well-being. Agriculture helps people prosper around the world and combines the creativity, imagination, and skill involved in planting crops and raising animals with modern production methods and new technologies. This series includes two main topics: Agronomy and Horticulture, and Animal Farming. This series will help readers better understand the intricacies of production agriculture and provide the new knowledge that is required to be successful. The success of a farmer in modern agriculture requires knowledge of events happening locally as well as globally that impact input decisions and ultimately determine net profit.

# Meet the Series Editor

W. James Grichar has been employed with Texas A&M AgriLife Research for over 45 years with an emphasis on research in agronomy, plant pathology, and weed science. He obtained his BS from Texas A&M in 1972 and his Masters of Plant Protection in 1975. He has published 195 journal articles, over 330 research reports and briefs, 11 book chapters, and over 300 abstracts of profession meetings. He also directs research in many crops including corn, grain sorghum, peanuts, and sesame. He has held various positions in different professional societies including the American Peanut Research and Education Society, Southern Weed Science Society, and Texas Plant Protection Conference in addition to being Associate Editor for Peanut Science and Weed Technology. Significant accomplishments have included spearheading efforts to determine the optimum planting time for soybean production along the upper Texas Gulf Coast. These efforts have shown growers that soybean yields can be improved by 10 to 20% by following a late March to early April plant date. He also has been instrumental in developing a herbicide program for peanut production in the south Texas growing region. Through the development and use of herbicides that are effective against major weed problems in the south Texas region, peanut yields have increased by 25 to 30%.

# Meet the Volume Editor

Dr. Ganesh Chandrakant Nikalje is an Assistant Professor (Senior Scale) of Botany specializing in plant biotechnology, stress physiology, and sustainable crop improvement. He completed his Ph.D. at Savitribai Phule Pune University, in collaboration with the Bhabha Atomic Research Centre, Mumbai, and pursued postdoctoral research at Gyeongsang National University, South Korea, under the SERB-SIRE program. His research focuses on molecular and metabolomic mechanisms of abiotic stress tolerance, halophyte biology, and climate-resilient crops, including physiological and molecular responses of soybean under salinity stress. Dr. Nikalje has published extensively in reputed international journals, edited several scientific books, and serves as a reviewer and editorial board member for leading plant science journals. He has received multiple national awards recognizing his contributions to plant stress biology and sustainable agriculture research.

# Contents

# Preface

Soybean (*Glycine max (L.) Merrill*) has emerged as one of the most significant global crops due to its multifaceted role in ensuring food security, nutritional sustainability, industrial applications, and climate-resilient agriculture. Increasing global demand for plant-based protein, edible oil, and sustainable agricultural systems has intensified the need for innovative scientific approaches that integrate agronomy, stress biology, digital technologies, and socioeconomic perspectives. The edited volume *Soybean Science and Sustainability – Productivity, Climate Resilience, and Smart Agriculture* brings together contemporary research and interdisciplinary insights to strengthen soybean production systems under changing environmental and agricultural conditions.

This volume is organized into four thematic sections that collectively address foundational production practices, climate adaptation mechanisms, emerging digital technologies, and farmer-centered adoption perspectives. The chapters included in this book provide a balanced integration of theoretical understanding, applied research, and technological advancement, offering valuable resources for researchers, academicians, students, policymakers, and agricultural practitioners.

Section I: *Foundations of Sustainable Soybean Production* begins with the chapter, *Agronomic Practices for Soybean Production under Climate-Resilient Conditions*, which discusses adaptive cultivation strategies to maintain productivity under variable climatic conditions. The chapter emphasizes sustainable crop management practices, improved resource-use efficiency, and agronomic interventions that enhance resilience against environmental uncertainties.

The next chapter, *Effective Nutrient Management Strategies for Soybean Production*, highlights the importance of balanced fertilization, soil health management, and nutrient-use optimization in achieving sustainable soybean yields. It examines integrated nutrient management approaches that combine organic and inorganic inputs to improve productivity while maintaining long-term soil fertility.

The section concludes with *Competitive Dynamics and Yield Performance of Soybean-Based Intercropping Systems on Inceptisols*, which explores intercropping as a sustainable agricultural strategy. The chapter evaluates crop interactions, resource competition, and the productivity advantages of diversified cropping systems, demonstrating how intercropping enhances land-use efficiency and ecological sustainability.

Section II: *Climate Adaptation and Stress Tolerance* focuses on soybean responses to abiotic stress conditions. The chapter, *Wild and Cultivated Soybean: Mechanisms of Salt Tolerance and Agricultural Applications*, provides insights into the physiological, biochemical, and molecular mechanisms underlying salinity tolerance. By comparing wild and cultivated soybean genotypes, this contribution highlights genetic resources and adaptive traits that can be used to breed climate-resilient soybean varieties suitable for marginal and saline environments.

Section III: *Digital and Precision Agriculture Approaches* presents emerging technologies transforming soybean cultivation. The chapter *Advancing Soybean Yield Prediction through Remote Sensing: A Review of Spectral Imaging Applications* reviews the application of remote sensing technologies and spectral analysis in monitoring crop growth, assessing stress responses, and predicting yield outcomes. These approaches enable large-scale, non-destructive crop assessment and support data-driven agricultural decision-making.

Complementing this perspective, *Advanced Machine Learning Approaches for Soybean Yield Prediction in Uttarakhand* demonstrates the application of artificial intelligence and machine learning models for accurate yield forecasting. The chapter illustrates how computational tools can enhance predictive accuracy, optimize farm management strategies, and contribute to precision agriculture frameworks.

Section IV: *Socio-economic and Adoption Perspectives* expands the discussion beyond biological and technological dimensions. The chapter, *Farmers' Responses to Soybean (*Glycine max *(*L.*)* Merrill*) Farming and the Influencing Socio-economic Factors in Jember District, Indonesia*, examines farmers' perceptions, adoption behavior, and the socioeconomic determinants influencing soybean cultivation. This contribution underscores the importance of integrating technological innovations with local knowledge systems and policy support to ensure the successful implementation of sustainable agricultural practices.

Collectively, the chapters in this edited volume highlight the transition of soybean research from conventional production approaches toward integrated, climate-smart, and technology-driven agricultural systems. The book emphasizes sustainability not only in terms of productivity enhancement but also through environmental conservation, efficient resource management, and socio-economic inclusiveness.

The editor sincerely acknowledges all contributing authors for their scholarly efforts, scientific rigor, and timely cooperation throughout the preparation of this volume. Appreciation is also extended to the reviewers for their valuable suggestions that helped improve the quality and clarity of the chapters. Gratitude is expressed to the publishing team for their professional guidance and support during the editorial and production processes.

It is hoped that this book will serve as a comprehensive reference for advancing soybean research and sustainable agricultural development, while also inspiring future innovations in crop science, precision agriculture, and climate-resilient farming systems.

**Ganesh Chandrakant Nikalje**
Department of Botany,
Seva Sadan's R. K. Talreja College of Arts, Science and Commerce,
Thane, India

Section 1

# Foundations of Sustainable Soybean Production

Chapter 1

# Agronomic Practices for Soybean Production under Climate-Resilient Conditions

*Tayyiba Usman and Muhammad Talha Ramzan*

## Abstract

Soybean is a crucial legume crop that significantly contributes to global food security and agricultural economies. However, climate change—characterized by temperature fluctuations, irregular precipitation, and increased pest and disease outbreaks—poses substantial threats to soybean productivity. These challenges necessitate the development of agronomic strategies that ensure stable yields under climate-resilient conditions. This chapter employed a comprehensive analysis of agronomic practices tailored to mitigate the impacts of climate stress on soybean production. Chapter methodology focused on selecting climate-resilient soybean varieties, determining optimal sowing times, and implementing conservation tillage, integrated nutrient management, and efficient water use. Additionally, practices such as intercropping and crop rotation with maize were evaluated for their yield-enhancing potential. This chapter concluded that combining these agronomic techniques significantly improves soil health, water efficiency, and plant resilience, ultimately leading to enhanced soybean yields even under adverse climate conditions. The practical benefits of adapting and integrating targeted farming strategies to build resilience against environmental stresses. The primary aim of this chapter was to identify and recommend a suite of sustainable agronomic practices that support soybean productivity in the face of climate variability, thereby contributing to long-term agricultural sustainability and food security.

**Keywords:** soybean production, agronomic practices, climate resilient, conservation tillage, intercropping, crop rotation, soil health

## 1. Introduction

Soybean (*Glycine max*) stands as one of the world's most valuable leguminous crops, playing a critical role in food security, livestock feed, and industrial applications due to its high protein and oil content. Over the decades, global demand for soybean has steadily increased, driven by population growth, urbanization, and dietary shifts. In this context, soybean production is pivotal to the agricultural economies of many nations [1]. However, climate change poses a growing challenge to the sustainable cultivation of soybean. Variations in temperature, erratic rainfall, and heightened incidences of pests and diseases threaten productivity and the stability of soybean yields [2]. These emerging climate-related stressors necessitate the adoption of agronomic practices that are both adaptive and resilient to changing environmental conditions. Temperature fluctuations, especially heat stress during the reproductive phase, significantly reduce pod setting and grain filling, leading to lower yields [3].

On the same note, any shift in precipitation whether in the form of long drought or surges of rainfalls affects the soil moisture level and plant growth, which results in the loss of productivity. The climate change has also changed the dynamics of pests and diseases where warmer environments and new levels of humidity modify the frequency and virulence of soybean pathogens and insect pests [4]. Such alterations indicate the necessity to adopt climate-resilient practices to preserve the potential of the crop. To reduce these risks and enhance the soybeans to perform well in varies climatic conditions, agronomists have resorted to several adaptive strategies. Selecting climate-resilient soybean varieties is crucial for ensuring stable crop performance. Such varieties can better withstand both pre-harvest and post-harvest biotic and abiotic stresses that may otherwise reduce yield. In addition, adjusting sowing times and adopting effective seed management practices enhance the chances of successful crop establishment and allow soybeans to make the most of their growing season [5].

Conservation tillage has the benefits of preserving structural integrity and water content in the soil, and integrated nutrient management (INM) provides a healthy and sustainable nutrient balance' effective water management measures such as drip irrigation and mulching play a critical role in both combating water shortage and maximizing the efficiency of use [6]. Besides agronomic interventions used independently, such interventions as intercropping and crop rotation, especially of maize, have exhibited a synergistic effect in increasing the yield production of soybean and maintaining soils fertility. These methods not only enhance the efficiency of land use but also lessen the accumulation of pests and diseases aiding to sustainable production system. This chapter aims to explore and recommend agronomic practices that enhance soybean productivity under climate-resilient conditions.

## 2. Impact of climate change on soybean production

### 2.1 Temperature stress

Changing climate of the world that is causing changes in temperature and rainfall patterns has raised a concern regarding to soybean crop yield. The minimum temperature for soybean is 17–18°C; below that, this crop cannot perform its physiological functions (like plant growth and root shoot formation) properly; likewise, maximum/optimum temperature for soybean crop is 30°C; elevated temperatures during reproductive stages, especially >35°C, significantly reduce yield potential [7]. Temperature stress affects soybeans at every stage of growth. Low and high temperatures impact the crop differently at various stages, as each growth stage has specific requirements for nutrients, water, and temperature. High-temperature stress, in particular, can lead to root and shoot distortion, poor seed germination, and disruption of physiological and biochemical processes in soybean plants. Review indicates that soybeans can generally tolerate temperatures between 32–38°C during the growing season. However, exposure to temperatures above 40°C during the flowering stage can damage flowers and young pods, resulting in fewer seeds, reduced seed weight, and ultimately lower crop yield [8].

Soybean seeds can germinate at temperature between 5 and 40°C, but with prolonged exposed to high temperature, their germination potential decreased; soybean seeds that were exposed to temperature of 27/22°C, day/night showed 84% of germination percentage, while those exposed to temperature of 32/28°C, day/night for 35 to 60 days showed germination percentage of 73 and 56%, respectively; hence, high temperature stress can be problematic at seed germination to flowering stage and low temperature stress can limit growth and development processes; it can disturb growth by affecting the physiological processes like nutrient and water absorption and protein and cellular metabolism [9]. Thus, temperature has a lot of effect on soybean growth and yield, so we must adopt some agronomic techniques like change in sowing time to prevail harsh temperature effects and need-based provision of all necessary components like water, nutrients, and so on. Using these approaches, we can make microbiome (the diverse community of microorganisms inhabiting soil) that can help the crop to withstand changing conditions and produce good yield [10].

### 2.2 Altered precipitation patterns

Climate change is altering the precipitation patterns worldwide; these changes are affecting crops production and the growth and quality of soybean crop. In several regions of the world, precipitation levels fall below the normal range, resulting in limited water availability for crop production. Since precipitation plays a vital role in meeting the moisture requirements of soybean, its deficiency can severely affect crop performance. Insufficient rainfall may lead to poor seed germination, weak seedling establishment, and reduced physiological activity, ultimately lowering the overall quality and productivity of the crop [11]. If there is excessive precipitation, then it also has negative effects on soybean crop resulting in suffocation of soybean crop that leads to lodging. Soybeans have a shallow root system that makes it highly sensitive to waterlogging and drought. Shifts in monsoon timing and intensity in Pakistan exacerbate these risks. So, we must adopt some site-specific agronomic measures like drip irrigation; by this method, if there is less or more precipitation, it will have minimum effect on crop growth because every plant is getting its water need accordingly [12].

### 2.3 Pest and disease dynamics

Climate change is causing a serious impact on soybean yield and quality by affecting the insects and disease attack patterns. Insect pests are more dependent on climatic factors like temperature, light, and precipitation, and changes in the pattern of these things can alter the biological characters and ecology of insects that is also affecting their damage and infection rate in soybean [13].

*Impact of insect pest on yield of soybean:* First thing of concern in this respect is the changing behavioral characteristics of these insects that is also changing their range of distribution and increase in their population density that can cause introduction of new insects to some areas or more virulent nature of existing ones. Secondly, climate change can also influence the transfer pathways and speed of their transmission or transmission of disease, changing their season of growth due to change in temperature either increasing or decreasing [14]. Changes in precipitation patterns significantly influence pest and pathogen dynamics in soybean cultivation. Excessive rainfall and increased water flow create favorable conditions for the proliferation and transmission of insect pests and disease-causing organisms. Climate change, therefore, alters the nature and intensity of insect attacks, ultimately reducing crop yield.

For instance, infestation by pod borers damages developing pods by hollowing them out, which leads to a decline in both seed quantity and quality. Such pest-induced stress, occurring at critical growth stages, poses a major constraint to sustainable soybean production under changing climatic conditions [15].

*Impact of insect pest on quality of soybean:* Insects decrease crop quality of soybean like by infesting it at different stages; as discussed before, it can affect the growth by infesting various parts, reducing the vigor of the crop, and ultimately, it can produce the crop but not of good quality like less crop stand and shrunk or immature seed and so on. Higher humidity and temperature favor the spread of pests such as stem fly and pod borers. Rust and bacterial blight have become more common due to erratic weather [16].

## 3. Selection of climate-resilient soybean varieties

### 3.1 Optimal sowing time and seed management

Soybean is very nutritious and important crop in this context; if its quality is reduced because of it being affected by climatic extreme or changing climatic conditions, then it is alarming for us to produce a good crop stand that can fulfill our daily needs; we must adopt some climate-resilient techniques to encompass climate change effects [17]. Changing sowing time, shifting it to 15 days before the already adopted sowing time, or moving it to 15 days ahead can improve our soybean crop production by mitigating effects of all environmental changes; this shifting can has positive effects on vegetative and reproductive stages of the crop as we have discussed earlier that high or low temperature can affect vegetative and reproductive stage of the crop; by this technique, we can save it from temperature effects; it can also help in fulfilling water requirements of the crop if we sow before 15 days of prescribed sowing time; then, the crop can take advantage of monsoon rains and be saved from drought stress that can retard its growth. Sowing between June 20 and July 10 maximizes crop establishment and avoids terminal drought [17]. For establishing a strong crop stand, the use of high-quality seed and provision of an appropriate growing environment are essential. A good seed is characterized by a high germination rate, strong vigor against insect and disease pressures, and efficient water use. Just as a solid foundation is crucial for constructing a building, high seed quality forms the basis for successful crop production. Seed vigor can be further enhanced through seed treatments with bio-inoculants, which protect seedlings from soil-borne pathogens and pests until germination. For example, inoculation with *Bradyrhizobium japonicum* not only improves seedling resistance but also enhances the biological nitrogen fixation capacity of soybean, thereby supporting better crop establishment and growth [18].

### 3.2 Conservation tillage and soil health

Conservation tillage is a sustainable agricultural practice aimed at improving soil health while conserving soil and water resources. It involves crop sowing with minimal soil disturbance, thereby reducing erosion and preserving soil structure. As part of conservation agriculture, complementary practices such as zero tillage (no-till farming) have gained recognition as reliable approaches to sustainable crop production. Compared to conventional tillage systems, no-till practices significantly enhance soil properties, particularly by increasing soil organic carbon content by up to 30%, which contributes to improved soil fertility, water retention, and long-term agricultural sustainability [19]. Conservation agriculture is a method of managing agroecosystem to improve sustainability and ensure food security by following the principles of minimum soil disturbance, permanent soil cover, and crop diversification.

Conservation tillage is any tillage system that leaves at least 30% of the soil surface covered with crop residue after planting to reduce soil erosion by water [20]. So, it can improve soil physical properties like it maintains more soil moisture in upper layers of soil than other farming techniques like plowing, so it can improve water use efficiency of the soil; it can also improve chemical properties of soil as it can preserve pH value and exchangeable cations of the soil that are disturbed by various tillage operations and give a better environment to crop growth [20]. Zero tillage has shown to improve water infiltration and reduce evaporation losses. Adoption in Multan and Faisalabad has improved soil organic matter and reduced production costs. Conservation tillage operations not only improve soil health but also increase soil organic matter and infiltration rate that improve fertilizer utilization of the crops, and it increases the yield of the crops. In most countries of the world, conservation or no-till system is being used; studies show that there is increased crop production using this; moreover, it improves soil health and crop environment by reducing the emission of greenhouse gases that cause environmental degradation [21].

### 3.3 Integrated nutrient management (INM)

Rapidly increasing global population, there is an urgent need to enhance crop production by 50–70% in order to meet future food security demands. Achieving this goal requires the cultivation of crop varieties with greater yield potential while adopting practices that ensure sustainable agriculture in the face of climatic extremes. One of the most effective approaches to improving both soil productivity and environmental sustainability is Integrated Nutrient Management (INM). This practice combines traditional and modern nutrient management strategies to optimize the use of organic and inorganic inputs, thereby enhancing crop quality and yield while minimizing environmental degradation. INM encompasses efficient methods of nutrient application, conservation practices, and the use of innovative technologies to improve the nutrient availability to plants. Furthermore, it emphasizes the dissemination of knowledge among farmers and researchers to ensure the widespread adoption of sustainable nutrient management practices. Earlier, nutrient management approach was only used to improve crop production, but now, along with improving production, it involves soil conservation [22].

Nutrient management has two major concepts including IPNS and INM (integrated plant nutrient supply system) that involve complete regulation of nutrients supply within the plants and other INM (Integrated nutrient management). Integrated Nutrient Management (INM) encompasses all aspects of soil and crop management, including the balanced supply of macro- and micronutrients, optimization of fertilizer use efficiency, selection of appropriate fertilizer types, and consideration of soil characteristics and nutrient requirements. The core principle of INM is to provide nutrients in forms that are most accessible and acceptable to plants, thereby ensuring efficient uptake and utilization. In soybean cultivation, INM practices play a crucial role in enhancing crop resilience to abiotic stresses such as heat and water stress. The application of organic amendments, including farmyard manure and compost, not only improves soil fertility but also stimulates microbial activity, strengthens soil buffering capacity, and enhances plant tolerance to stress, ultimately contributing to sustainable crop productivity (**Figure 1**) [23].

**Figure 1.**
*Agronomic practices for enhancing soybean production.*

### 3.4 Efficient water management

Decreasing groundwater table is the matter of high concern for agriculture as most of the underground water is being used for agricultural irrigation that is now decreasing. Due to global warming, hydrological cycle is being disturbed due to elevated temperature causing more evaporation than the normal one, and water being applied as irrigation is not completely reaching to crops due to leaching processes, and rainfall patterns are also disturbed causing more rains in tropical areas and comparatively drought condition in semiarid areas [24]. Water productivity in the agriculture can be achieved by integrating water management practices together with adapting techniques of precise irrigation and soil health safeguarding approaches. In addition to climate change, several other factors significantly influence crop production. These include the productivity and fertility of soils, the prevalence of crop diseases, the intensity of pest infestations, and the selection of appropriate crop varieties. Each of these factors interacts with environmental conditions and management practices, collectively determining the overall yield and quality of agricultural production [25].

Water management can be done by adopting various techniques like water storage, underground water channels, and metered application of water to reduce evaporation and leaching losses; certain water management practices can reduce water use by 30–40%. As soybean crop has specific water requirements and is a sensitive crop, its production can be depleted if its moisture level cannot fulfill its evapotranspiration rate. There are various methods that can be used to use irrigation water efficiently like subsurface irrigation systems like drip irrigation or micro irrigation [24].

Water management through micro-irrigation has proven to be an effective strategy for enhancing both crop yield and the quality of agricultural produce. By delivering water directly to the root zone, micro-irrigation systems ensure efficient utilization of water resources while minimizing losses through evaporation, runoff, and deep percolation. This targeted application not only contributes to substantial water savings but also helps in mitigating problems such as soil salinity and excessive nitrate

| Practices | Application | Benefits | References |
|---|---|---|---|
| Selection of Climate Resilient Soybean Varieties | Use of drought, heat, and disease-resistant soybean varieties such as JS 95-60, MAUS 162, RVS 2001-4, NRC 127, DSB 21 | Withstands climate stress (drought, heat). Reduces yield losses. Increases adaptability | [26] |
| Optimal Sowing Time and Seed Management | Sowing during the optimal window (e.g., late June to early July in India); using treated and high-vigor seeds | Maximizes yield potential. Reduces vulnerability to early/late-season drought and pest attacks | [27] |
| Conservation Tillage and Soil Health | Minimum tillage, retention of crop residues, and cover cropping | Enhances soil moisture retention. Reduces erosion. Improves organic matter and microbial activity | [28] |
| Integrated Nutrient Management (INM) | Balanced application of organic manures (FYM, compost), biofertilizers (Rhizobium, PSB), and chemical fertilizers based on soil testing | Optimizes nutrient availability. Improves soil fertility. Reduces input cost and environmental risk | [29] |
| Efficient Water Management | Use of mulching, alternate furrow irrigation, and rainwater harvesting; adoption of micro-irrigation systems (drip/sprinkler) | Conserves water. Increases water-use efficiency. Reduces crop water stress | [30] |

**Table 1.**
*Applications and benefits of various climate-resilient soybean farming practices.*

leaching. Consequently, micro-irrigation supports sustainable crop production by improving resource use efficiency and maintaining soil and environmental health. It can also be done by efficiently scheduling of irrigation water as it should be applied at peak time of crop need and should be saved for future use by using water harvesting techniques. Water conservation practices can also be used to save the moisture of soil from being evaporated so that crop can get adequate moisture and can be saved from climatic abnormalities occurring due to climate change (**Table 1**) [24].

## 4. Agronomic practices to enhance soybean yield

### 4.1 Intercropping with maize enhance soybean yield

Maize-soybean rotation method improves soil carbon and nitrogen retention when soybeans are grown. According to recent studies, maize yields in a rotation system that alternates maize with soybeans for 2 and 3 years were 9.4 and 12.6% higher, respectively, than those in a system that alternates maize with sweet potatoes [31]. According to the data, intercropping soybean-based maize strips improves soil fertility and yields more of the subsequent crop by raising phosphate availability and APase activity. An annual alternate rotation between maize and soybean strips has been employed in maize-soybean strip intercropping systems to boost the farmland's sustainable output and avoid soil continuous cropping barriers [32].

Annual rotation between maize and soybean strip alternative crops increased maize grain yield and absorption amount of nitrogen, phosphorus, and potassium in maize by 7.5, 18.5, 9.1, and 14.2%, respectively, compared to successive maize-soybean strip intercropping systems, without significantly altering soybean yield. As an alternative, existing strip intercropping systems balance high crop productivity and sustainability by obtaining additional soybean production without compromising maize yield because strip-intercropping is both economical and sustainable [33]. First off, this approach improves land-use efficiency, which is crucial for developing nations with sizable people and little forested area. In many poor nations, the implementation of maize/soybean rotation systems is nearly impossible due to land constraints. Thus, it is important to note that the intercropping system of maize and soybean strips would maximize sustainable land productivity.

Second, this extremely effective intercropping technology can generate a lot more food for both humans and animals. Third, labor productivity would significantly rise with the use of agricultural equipment suitable for this approach. Fourth, soybeans' capacity to fix nitrogen would increase soil fertility and provide other crops with more organic fertilizer, both of which would significantly lower pollution [34]. According to FAO statistics from 2009 to 2013, developing Asian countries' croplands are covered by roughly 56.53 million hectares of maize and 19.12 million hectares of soybeans annually, while African countries' croplands are covered by roughly 33.13 million hectares of maize and 1.52 million hectares of soybeans. More output would be produced each year if this maize-soybean strip intercropping technique was used in all these areas, suggesting that it has a lot of potential to keep agricultural sustainability in developing nations and avert food crises [35].

### 4.2 Crop rotation with maize enhance soybean yield

Using soy-maize rotations in the tropics has several agronomic advantages, such as improved soil fertility, less biotic pressure, and higher yields of soy and maize each year. Comparing soy-maize rotation to cereals grown continuously, SSA cereal yields in areas planted following a legume are increased by an average of 0.49 tons/hectare or higher [36]. When compared to continuous cereal production, soy-maize rotations are also highly beneficial and sustainable over time since they can sustain high levels of agricultural productivity after many years of agriculture. Several studies show that soy rotation outperforms intercropping systems in terms of yield increase. Farmers have historically grown maize either constantly or in combination with other crops, such pulses, in an intercropping strategy [37].

In intercropping systems, many crops are grown on the same piece of land in the same season. Although intercropping can help diversify household diets, in certain situations, crop competition can be detrimental and impact yields. Crop rotation involves planting a different monocrop one after the other throughout each growing season. For instance, in Brazil's central west, where the rainy season might extend 6 months, maize and soybeans coexist in the "safrinha" system [37]. The capacity of soy to fix atmospheric nitrogen into the soil makes soy rotations a popular sustainable approach to nutrient management in agricultural settings. Aboveground soybean leftovers are high in nitrogen and can be added to the crop the next year or used as animal feed. This in turn feeds a cycle whereby high-quality leftovers produce high-quality manure, which may then be applied to the soil to further increase its fertility. Subterranean soybean leftovers added to the soil after harvest also increase its organic matter and nitrogen content [36].

Soybean rotations reduce the requirement for chemical fertilizer inputs by fixing 17–450 kg·ha$^{-1}$ of nitrogen annually in tropical environments, and then releasing

| Practice | Application | Benefits | References |
|---|---|---|---|
| Intercropping with Maize | Plant soybean and maize simultaneously in alternating rows (e.g., 1:1 or 2:1 ratio). Maintain proper spacing to reduce competition | Improves land use efficiency. Enhances soil fertility through nitrogen fixation by soybean. Reduces pest and disease pressure. Increases income per unit area | [38] |
| Crop Rotation with Maize | Rotate soybean with maize seasonally or annually (e.g., maize in Year 1, soybean in Year 2). Use legumes to break cereal cycles | Breaks pest and disease cycles. Improves soil fertility and organic matter. Enhances nitrogen availability. Increases long-term yield stability | [37] |

**Table 2.**
*Agronomic practices (intercropping and crop rotation with maize) to enhance soybean yield.*

0–170 kg·ha$^{-1}$ of accessible nitrogen for the subsequent maize crop. This is rather beneficial to the smallholder farmers, most of whom are unable to afford fertilizers. The long-term yield of maize could be increased because of this increase in the amount of nitrogen in the soil because of soy rotations. Moreover, due to soybeans producing root exudates, rotations were able to increase the supply of phosphorus from soil and the phosphate rock fertilizer to the cereal crops in phosphorus-deficient conditions (**Table 2**) [39].

## 5. Conclusion and future prospective

Agronomic practices that are climate resilient are important in sustaining and increasing soybean yield under the growing dynamics of climate variability. This chapter indicates that an increase in temperature stress, changes in precipitation patterns, and shifting pest and disease dynamics have a profound impact on soybean growth and development as well as yield. A combination of the strategies inclusive of choosing climate resilient varieties of soybean, the timely sowing (sowing-time) of beans, the use of conservation tillage, adopting integrated nutrient management (INM), and effective use of water has worked to combat all these effects of climatic changes. Several agronomic practices have been employed to enhance both crop productivity and environmental sustainability. Among these, intercropping and crop rotation, particularly with maize, play a significant role. These practices not only improve overall agricultural production but also contribute to enhanced soil fertility, better nutrient cycling, and greater ecosystem sustainability. By diversifying cropping systems, such measures reduce pest and disease pressure, improve resource-use efficiency, and strengthen the resilience of farming systems under changing climatic conditions. The results indicate the necessity of location-specific, dynamic management strategies capable of resisting soybean farming against climatic calamities as well as supporting sustainable increases in agricultural intensification.

Development of region-specific soybean genotypes capable of controlling various stress generated by climate should be the priority of the future studies. More attention should also be directed to incorporation of digital agriculture technologies, including remote sensing, AI-based modeling, and precision farming to optimize decision-making on the farm. In addition, promotion of the resistance practices will require the improvement of awareness of farmers using extension services and participatory research. It will also be important to enhance the climate-resilient agriculture policies, as well as the incentives toward sustainable land and water management. With climate change increasingly becoming a multidimensional challenge to food security, we expect a comprehensive, multidisciplinary intervention based on agronomy, technological, and policy interventions to become the only way to ascertain sustainability of soybean production and sustainable environment.

## Authors contributions

All authors have equal contributions.

## Conflict of interest

The authors declare that they do not have any personal relationships that can affect the work reported in this Chapter.

## Author details

Tayyiba Usman* and Muhammad Talha Ramzan
Department of Agronomy, University of Agriculture Faisalabad, Punjab, Pakistan

*Address all correspondence to: tayyibausman197@gmail.com

## References

[1] Vishnoi NK. An overview on soybean. Asian Journal of Multidimensional Research. 2021;**10**(11):205-210

[2] Alı W et al. Nutritive potentials of soybean and its significance for humans health and animal production: A review. Eurasian Journal of Food Science and Technology. 2020;**4**(1):41-53

[3] Jithender B, Nickhil C. Spotlighting Soybeans in Food: Diet and Health Promoting Characteristics. Vol. 2, No. 4. Indian Food Industry Mag; Jul–Aug 2020

[4] Toomer OT et al. Current agronomic practices, harvest & post-harvest processing of soybeans (Glycine max)—A review. Agronomy. 2023;**13**(2):427

[5] Zhong Y, Zhong S. Advances in agronomic practices for high-yield soybean cultivation. Bioscience Methods. 2024;**15**(1)

[6] Nehbandani A et al. Comprehensive yield gap analysis and optimizing agronomy practices of soybean in Iran. The Journal of Agricultural Science. 2020;**158**(8-9):739-747

[7] Jumrani K, Bhatia VS. Interactive effect of temperature and water stress on physiological and biochemical processes in soybean. Physiology and Molecular Biology of Plants. 2019;**25**(3):667-681

[8] Jumrani K, Bhatia VS. Impact of combined stress of high temperature and water deficit on growth and seed yield of soybean. Physiology and Molecular Biology of Plants. 2018;**24**(1):37-50

[9] Mederski HJ. Effects of water and temperature stress on soybean plant growth and yield in humid, temperate climates. In: Crop Reactions to Water and Temperature Stresses in Humid, Temperate Climates. Boca Raton (FL): CRC Press; 2019. pp. 35-48

[10] Ogunkanmi L, MacCarthy DS, Adiku SG. Impact of extreme temperature and soil water stress on the growth and yield of soybean (*Glycine max* (L.) Merrill). Agriculture. 2021;**12**(1):43

[11] dos Santos CA et al. Trends of extreme air temperature and precipitation and their impact on corn and soybean yields in Nebraska, USA. Theoretical and Applied Climatology. 2022;**147**(3):1379-1399

[12] Bijlwan A, Ranjan R, Kumar M, Jha A. Climate change: Projections and its possible impact on soybean. In: Soybean Production Technology: Physiology, Production and Processing. Singapore: Springer; 2025. pp. 19-44

[13] Lee ST, Li C, Davis JA. Predator-pest dynamics of arthropods residing in Louisiana soybean agroecosystems. Insects. 2022;**13**(2):154

[14] Kalyan R, Ameta O. Effect of sowing time and varieties on incidence of insect pests of soybean. Journal of Entomology and Zoology Studies. 2017;**5**(2):790-794

[15] Suyal P et al. Seasonal incidence of insect pests and their natural enemies on soybean crop. Journal of Entomology and Zoology Studies. 2018;**6**(4):1237-1240

[16] Hesler LS et al. Early-season pests of soybean in the United States and factors that affect their risk of infestation. Journal of Integrated Pest Management. 2018;**9**(1):19

[17] Jarecki W, Bobrecka-Jamro D. Effect of sowing date on the yield and seed quality of soybean (*Glycine max (L.)* Merr.). Journal of Elementology. 2021;**26**(1):85-96

[18] Kundu PK et al. Effect of sowing date on yield and seed quality of soybean. Journal of Agriculture and Ecology Research International. 2016;**9**(4):1-7

[19] Kumar S, Garg AK, Aulakh MS. Effect of conservation agriculture practices on physical, chemical and biological attributes of soil health under soybean–rapeseed rotation. Agricultural Research. 2016;**5**:145-161

[20] Sangotayo AO et al. Long-term conservation and conventional tillage systems impact physical and biochemical soil health indicators in a corn–soybean rotation. Soil Science Society of America Journal. 2023;**87**(5):1056-1071

[21] Wulanningtyas HS et al. A cover crop and no-tillage system for enhancing soil health by increasing soil organic matter in soybean cultivation. Soil and Tillage Research. 2021;**205**:104749

[22] Singh N, Kushwaha H, Singh A. Integrated nutrient management on growth, yield, nutrient uptake and fertility balance in soybean (*Glycine max L.*)-wheat (*Triticum aestivum L.*) cropping sequence. International Journal of Bio-resource and Stress Management. 2020;**11**(4):405-413

[23] Bonde AS, Gawande S. Effect of integrated nutrient management on growth, yield and nutrient uptake by soybean (*Glycine max*). Annals of Plant and Soil Research. 2017;**17**(2):154-158

[24] Aminah A et al. Effectiveness of water management towards soil moisture preservation on soybeans. International Journal of Agronomy. 2020;**2020**(1):8653472

[25] He J et al. Conserved water use improves the yield performance of soybean (*Glycine max* (*L.*) *Merr.*) under drought. Agricultural Water Management. 2017;**179**:236-245

[26] Sarkar J. Soybean production techniques. In: Soybean Production Technology: Physiology, Production and Processing. Vol. 2. Apple Academic Press (CRC Press/Taylor & Francis Group); 2025. p. 125

[27] Kumar D, Gurjar D, Sharma UP, Kumar D. Analysis of the Process of Seed Viability and Vigor Testing. New Delhi: Akinik Publications; 2021. p. 36

[28] Li Y et al. Residue retention promotes soil carbon accumulation in minimum tillage systems: Implications for conservation agriculture. Science of the Total Environment. 2020;**740**:140147

[29] Choudhary AK, Rajanna G, Kumar A. Integrated nutrient management: An integral component of ICM approach. In: Integrated Crop Management Practices for Enhancing Productivity, Resource-Use Efficiency, Soil Health and Livelihood Security. New Delhi: Today & Tomorrow's Printers and Publishers; 2018. pp. 33-39

[30] Ambomsa A. Role of drip irrigation system as increasing water use efficiency over furrow irrigation system. Academic Research Journal of Agricultural Science and Research. 2020;**8**(3):252-262

[31] Du J-b et al. Maize-soybean strip intercropping: Achieved a balance between high productivity and sustainability. Journal of Integrative Agriculture. 2018;**17**(4):747-754

[32] Wu Y et al. 2023 Soybean yield variations and the potential of intercropping to increase production in China. Field Crops Research. 2023;**291**:108771

[33] Raza MA et al. Maize/soybean strip intercropping produces higher crop yields and saves water under semi-arid conditions. Frontiers in Plant Science. 2022;**13**:1006720

[34] Iqbal N et al. Comparative analysis of maize–soybean strip intercropping systems: A review. Plant Production Science. 2019;**22**(2):131-142

[35] Sentelhas PC et al. The soybean yield gap in Brazil–magnitude, causes and possible solutions for sustainable production. The Journal of Agricultural Science. 2015;**153**(8):1394-1411

[36] Agomoh IV et al. Crop rotation enhances soybean yields and soil health indicators. Soil Science Society of America Journal. 2021;**85**(4):1185-1195

[37] Acevedo-Siaca L, Goldsmith PD. Soy-maize crop rotations in sub-Saharan Africa: A literature review. International Journal of Agronomy. 2020;**2020**(1):8833872

[38] Ali A et al. Inter-and intra-row and plant spacing impact on maize (*Zea mays L.*) growth and productivity: A review. International Journal of Advanced Science and Research. 2017;**2**(1):10-14

[39] Yuan M et al. Long-term corn–soybean rotation and soil fertilization: Impacts on yield and agronomic traits. Agronomy. 2022;**12**(10):2554

Chapter 2

# Effective Nutrient Management Strategies for Soybean Production

*Umesh Kumar Singh, Santosh Korav, Dhanpreet Kaur, Thangavel Palanivel Swarnam and Niharika Kodirekka*

## Abstract

The soybean is one of the world's most significant legume crops, primarily serving the majority of global protein and oil needs. The production of soybean crops is limited by the various constraints, including biotic and abiotic factors, as well as crop management practices. For soybean crops, 15 essential nutrients are required nitrogen, phosphorus, potassium, calcium, magnesium, and sulfur, which are needed in relatively large quantities, and micronutrients comprising iron, boron, zinc, cobalt, copper, manganese, molybdenum, nickel, and chlorine. These nutrients can be supplied to soybean plants either by incorporating them into the soil or applying a commercially available fertilizer as a foliar spray. The concentration of nutrients required by soybean crops varies, and deviations from the optimal range can lead to deficiency or toxicity. The concentration of nutrients in the soil solution, their chemical form, the mode by which the plant takes up the nutrients, their interactions with other soil nutrients, soil chemistry, and the method of fertilizer application determine the availability of nutrients to soybean crops. This review examines essential nutrients for sustainable soybean production, addressing the roles and functions of nutrients, required concentrations, and visual symptoms that appear during deficiency, and incorporates the findings of several studies. This study aims to provide practical guidance on effective nutrient management for soybean farmers and to help scholars review the literature on soybean nutrient management.

**Keywords:** nutrient management, soybean, essential nutrient, micronutrients, climate resilient

## 1. Introduction

Soybean (*Glycine max*) is one of the world's most valuable leguminous crops. It is essential for food security, livestock feed, and industrial use because of its high protein and oil content. Over the decades, global demand for soybeans has steadily increased, driven by population growth, urbanization, and dietary changes. As a result, soybean production is crucial to the agricultural economies of many countries [1]. It is cultivated on 127.60 million hectares and yields 364.07 million tons [2] of global production each year. Its versatility makes it a wonder crop, and no soybean by-product is left unused. This crop has a greater chance of replacing many oilseeds, which helps fill the gap in edible oil and protein-rich foods. Soybean is one of the world's oldest cultivated crops. In

India, many people are vegetarians, so a crop like soybean, with its high protein content and strong yield potential, has become especially important. Soybean protein currently receives more attention than any other protein source. It also provides a range of vitamins, calcium, phosphorus, and iron. These qualities make it highly suitable for human consumption. Soybean is used in food products and in fermented foods such as soy sauce and cheese. Nutritionally, it contains 37–48% protein and 16–21% oil, varying by variety and agronomic practices [2]. Because its protein content exceeds that of a typical non-vegetarian diet, it is also called "poor man's meat." The seed also supplies vitamins, minerals, and beneficial compounds such as isoflavones, which offer protection against age-related disorders, cardiovascular disease, osteoporosis, and cancers. Soybeans are processed for vegetable oil, feed, and culinary uses, and are made into products such as nutrella, tofu, nattu, and soy drinks. Consequently, soybeans are consumed worldwide, and demand has risen by 145% since 1990/91 [3], far higher than the increases of 31% for rice, 76% for corn, and 35% for wheat [3]. The demand for soybean production is expected to stay strong in the coming years, as soybeans are useful in many sectors of the food industry including animal feed, foods and drinks, personal-care products, dietary supplements, pharmaceuticals, and bio-materials like biofuels yet India's average soybean yield is only about 1 ton per hectare, whereas the worldwide average is 2–3 t per hectare [4]. Recent research has emphasized that managing nutrients well is essential for modern soybean cultivars to boost yields in both direct and indirect ways. The studies show that new cultivars take up far more nutrients than older ones, highlighting the importance of carefully applying nutrients after accurate soil fertility tests. Using a variety of nutrient sources and applying them at the right growth stages with suitable techniques can improve nutrient efficiency and support sustainable production. In this review, nutrient management for soybean is examined in depth, covering the roles of nutrients, their availability in soil and plants, and the best practices for optimizing nutrient use and crop performance.

## 2. Importance of soybeans in global agriculture

China is the origin place of soybeans, which serve as high-quality animal feed and a major source of protein for humans. Rising demand for soybeans stems from growing consumption and from the substantial dietary nutrients they contain. With more than 23,000 varieties in Asia, soybeans were first domesticated in China before spreading to the United States and Brazil; a concise chronicle of their global distribution can be found. The word "soybean" also refers to the bean used to make soy sauce. Soybean is a primary global source of vegetable oil and animal-protein feed; among dietary legumes, it comes only after peanuts for oil content (18–22%) and boasts the highest protein content (40–42%) of any similar crop [3]. Until the 1980s, the United States supplied more than half of the world's soybean harvest, but today Brazil and Argentina have joined the United States as the key producers. Together, the United States, Brazil, Argentina, China, and India account for over 92% of global output, making them the world's leading soybean exporters, while Africa has also begun producing the crop since the twentieth century [3]. Presently, Brazil, one of the traditional "big five" grain exporters alongside the United States, Canada, Australia, Argentina, and the European Union, is a heavyweight in the tropical food industry. Since 1990, Brazil has supplied about one-third of all soybean exports worldwide, positioning it as the second-largest producer after the United States, which crops soybeans on only 6% of its arable land yet still contributes 25% of the global supply [3]. High soybean yields depend on substantial nitrogen (N), which can be provided *via* chemical fertilizers or bio-fertilizers. A

**Figure 1.**
*Different growth stages of soybean plant.*

more cost-effective alternative to conventional N fertilization is biological nitrogen ($N_2$) fixation by symbiotic soil bacteria, particularly *Bradyrhizobium*, which supplies a sustainable nitrogen source. Biological nitrogen fixation works best when the right plant, rhizobia, symbiosis, and environmental conditions are present. Because soybeans are such a key crop, governments, universities and private people around the world are studying many aspects of soybean farming, and recent research has helped farmers use better inoculants that benefit the environment and improve agro-ecosystem sustainability [4]. Yet current soil-microbe stewardship and conservation plan still miss the essential role of these microbes in soil biodiversity, and a study of the economic value of soybean nitrogen fixation in Africa showed that smallholder farmers gained the most, illustrating how valuable soybeans are for fixing nitrogen in that region (**Figure 1**) [3]. They highlighted promiscuous soybean varieties and suggested alternatives that would improve small-holder farmers' chances of benefiting from biological nitrogen fixation, a particularly valuable advantage when inorganic fertilizer is scarce and soybean yields need to increase, noting that none of the 19 African countries that cultivate soybeans currently grow these promiscuous types, yet the economic gains from nitrogen-fixing promiscuous soybeans show how soil microbial diversity can enhance human well-being; several cases favor biological nitrogen fixation; for example, some inoculated cultivars did not exceed the yield of the uninoculated promiscuous varieties even at their peak production, and with these findings in mind, the authors called on India to consider selecting and breeding promiscuous soybean varieties, given the complex response of plants to inoculation, since legume crops like soybeans can form nodules with many rhizobial strains in soils deemed promiscuous (**Figure 1**).

## 3. Macronutrient induction in soybean crops

### 3.1 Nitrogen

Nitrogen is a vital nutrient for soybeans, forming part of chlorophyll and enzymes that control the plants' physiology, and it is essential for building amino acids, the basic blocks of proteins, while also supporting carbohydrate metabolism that encourages root growth and the uptake of other nutrients. Soybeans demand the most nitrogen of any nutrient; they need roughly 80 kg of nitrogen to produce 1000 kg of seed, and to reach higher yields, more nitrogen must be supplied, as the plants continue to take it up throughout both vegetative and reproductive stages [5].

### 3.2 Phosphorus

Phosphorus is a vital primary nutrient for soybean crops, essential for proper growth and development. It plays a key role in storing and transporting the energy produced during photosynthesis, which is then used for growth, development, and reproduction. As a core component of ATP molecules, it boosts the efficiency of other applied nutrients. Phosphorus fertilization affects soybean yield, protein and oil content, nitrogen fixation, root development, leaf area, and stress tolerance. Because phosphorus moves slowly in soil, plants can show deficiency symptoms when it is located too far from the root zone. It also reacts with nutrients such as aluminum, calcium, and iron to form orthophosphate, rendering the phosphorus unavailable to the plant [6]. To achieve a yield of one ton of soybean seeds, soybean plants require 25 kg of phosphorus in the form of $P_2O_5$.

### 3.3 Potassium

Potassium plays many key roles in plants, from keeping cells turgid and moving starch, water, and nutrients to aiding protein synthesis and starch build-up; it also helps plants cope with stress, disease, and pests while ensuring other nutrients are taken up in balance. As a mobile nutrient, potassium is essential for activating enzymes during nodulation, strongly influencing seed yield, weight, and protein level, though its effect on seed oil content is small. For instance, one soybean plant needs 53 kg of potassium in the form of $K_2O$ to yield one ton of soybean seeds, and potassium can be added to soybean crops as potassium chloride [60–62%] or potassium sulfide [50%]; to maximize seed production, growers are advised to apply 60–100 kg $K_2O$ per hectare. Potassium demand is lower in the early growth phase and rises during vegetative development, and during the pod-filling stage, the potassium stored in vegetative tissues is redirected into the seeds [5].

### 3.4 Calcium

Calcium is vital for cell-wall integrity and stability, serving as a counter-cation for organic and inorganic anions in vacuoles. Calcium deficiency is rare in most crops. Plants absorb calcium from the soil through their roots, after which it is carried *via* xylem vessels to the shoots. For soybeans, 28 kg of calcium is required to produce one ton of seed yield. Calcium influences soybean production both directly and indirectly: directly, it boosts nodulation and nitrogen fixation; indirectly, it helps keep soil pH balanced and improves the availability of other nutrients to the plant [7].

### 3.5 Magnesium

Magnesium is a component of chlorophyll and plays a key role in photosynthesis; it also aids plant energy metabolism and protein synthesis. Plants can acquire magnesium from inorganic rock, but soybean crops receive it mainly through dolomite. In soil, magnesium exists as $Mg^{2+}$—a base cation in the soil solution—and its availability is strongly influenced by other cations such as calcium, manganese, ammonium, and potassium. Soybean plants need about 22 kilograms of magnesium to produce one ton of seed yield [7].

### 3.6 Sulfur

Sulfur is an essential macronutrient that promotes growth, enhances productivity, and increases the oil content of soybean seeds. Because sulfur is a key component of amino acids, it supports oil synthesis. Soybean plants require 7–8 kg of sulfur to produce a ton of seed. Researchers are actively seeking ways to improve sulfur use efficiency in soybean production [6]. In practice, sulfur is typically applied together with nitrogen and potassium in a combined form, using fertilizers such as ammonium sulfate, potassium sulfate, and calcium sulfate (gypsum).

## 4. Essential micronutrient induction in soybean crop

### 4.1 Copper

Copper is an essential micronutrient for soybean growth and development. Although excess copper is toxic, it plays crucial roles in plant physiology and serves as a cofactor in many enzymatic reactions. It supports key processes such as respiration, photosynthesis, lignin formation, and carbohydrate metabolism. Although copper is typically bound to minerals in complex form, plants absorb it as $Cu^{2+}$ ions. Because copper hardly moves within plant tissues, its localized buildup can cause toxicity. Farmers usually supply copper to soybean crops as copper sulfate, although commercially available chelated products also deliver copper [6]. Animal manure contains ample copper, and when applied carefully, it can alleviate copper deficiency in soybean fields.

### 4.2 Manganese

Manganese is essential for redox reactions such as the electron transport chain, chlorophyll formation, and the operation of Photosystem II, and it serves as an activator for more than 35 distinct enzymes. In the soil, it is present as $Mn^{2+}$, $Mn^{3+}$, and $Mn^{4+}$, but plants absorb it primarily in the $Mn^{2+}$ form. Farmers apply manganese sulfate to soybean leaves to meet their requirements. The amount of manganese available to soybeans depends largely on tillage, organic-matter content, and water movement, so deficiency is more common in flood-prone or water-fluctuating zones [5]. The most reliable way to increase seed yield is to apply 5–22 kg Mn $ha^{-1}$ of $MnSO_4$ together with starter fertilizer at planting time.

### 4.3 Boron

Boron is an essential trace element involved in many physiological functions: it helps build cell walls, produce carbohydrates, form nodules, fix nitrogen, maintain membrane integrity, promote root growth, aid tissue differentiation, trigger pollen germination, and support general plant development. Each ton of soybean seed yield requires about 55 grams of boron. Plants absorb boron mainly as boric acid [$H_3BO_3$], which dissolves in the soil, and an important additional source comes from soil organic matter released by the microbial activity [5].

### 4.4 Cobalt

Cobalt is essential for plant growth and development; its benefits include promoting stem elongation, delaying senescence, enhancing drought tolerance in seeds, and extending coleoptiles—this last effect helps activate various enzymes. Plants take up cobalt as $Co^{2+}$, which exists in the soil solution and on cation-exchange sites. For leguminous species, cobalt is a critical trace element because it works synergistically during nodulation and nitrogen fixation [8]. The highest gains in shoot length, root length, leaf number, branch number, and both fresh and dry root and shoot weights occur when the application rate is 12 mg $kg^{-1}$.

### 4.5 Iron

Iron is a vital element for every living being, serving as a cofactor and building block for many enzymes. In plants, it helps make chlorophyll and powers photosynthesis. It also supports DNA production, breathing, a range of metabolic reactions, and functions as a prosthetic piece of some enzymes. However, most of the iron leached from the course-textured soil, and a comparable replication of the experiment was conducted in clay soil, where iron applied at a rate of 40 kg $ha^{-1}$ yielded the highest production [6].

### 4.6 Chlorine

Chlorine is an essential micronutrient that supports various physiological and metabolic functions—such as regulating stomatal osmosis, releasing oxygen during photosynthesis, and enhancing disease resistance and tolerance. In the soil solution, chlorine is present as the chloride ion ($Cl^-$) or as soluble salts like sodium chloride, potassium chloride, and calcium chloride. Soybeans rarely suffer from chlorine deficiency because they can tolerate low levels. Soil chlorine builds up through rainfall and potassium fertilizers; for instance, KCl supplies most of the chlorine that soybean plants use. However, in poorly drained soils or when potassium fertilizers are applied excessively, chlorine can become toxic. Leaf scorch—a common sign of soybean plants treated with KCl—correlates positively with chlorine concentrations in leaves and seeds, and negatively with seed weight and yield [5].

### 4.7 Molybdenum

Molybdenum is an essential nutrient for leguminous crops because it plays a crucial role in nitrogen fixation; certain enzymes—such as nitrate reductase, xanthine dehydrogenase, aldehyde oxidase, and sulfite oxidase—use this metal to carry out redox reactions. As a key component of nitrogenase, molybdenum is required for the conversion of nitrates to nitrites and for reducing atmospheric nitrogen to ammonia during the fixation process. In the soil, molybdenum exists as the molybdate ion, which moves through the soil by mass flow alongside transpiration; drought conditions can slow this movement, thereby worsening molybdenum deficiency [7]. The availability of molybdenum to crops is strongly influenced by soil pH, with higher pH levels increasing its availability, while excess sulfate and iron can also promote deficiency in the soil.

### 4.8 Zinc

Zinc is an essential micronutrient for plants because it is a component of proteins and performs functional, regulatory, and structural roles in enzymes, influencing gene expression and stabilizing RNA and DNA structures. It is found in soil solutions either as a free ion or bound to a low-molecular-weight compound. In its divalent form ($Zn^{2+}$), it is mobile only within the soil, and plants take it up by diffusion

[8]. Zinc bioavailability mainly depends on mineral solubility, the amount of zinc adsorbed onto clay and organic matter, and soil pH.

### 4.9 Nickel

Nickel, a newly identified micronutrient, is essential for rhizobial nodulation in soybean crops and activates the plant urease enzyme that hydrolyses urea; it exists in the soil solution as $Ni^{2+}$, which stays adsorbed to soil cation-exchange sites, but its uptake is suppressed when copper and zinc ions are present in high concentrations [9].

## 5. Nutrient availability and uptake mechanisms

### 5.1 Nitrogen (N) availability and uptake: The symbiotic advantage

Nitrogen is an essential nutrient for soybean growth and protein production, and the plant acquires it primarily through a distinct mechanism.

(a) *Primary mechanism: Biological nitrogen fixation (BNF)*

Soybeans partner with *Bradyrhizobium japonicum* in root nodules, where the bacteria convert atmospheric nitrogen ($N_2$) into plant-usable ammonium ($NH_4^+$). This process supplies about 50–75% of the plant's nitrogen needs, markedly reducing fertilizer dependence [6].

(b) *Secondary mechanism: Soil uptake*

Soybeans also absorb nitrate ($NO_3^-$) and ammonium directly from the soil. Elevated soil nitrate levels can inhibit nodule formation and BNF, forcing greater reliance on the less energy-intensive soil source.

(c) *Micronutrient link*

Efficient BNF depends on micronutrients molybdenum (Mo) and cobalt (Co), which are the critical components of the bacterial nitrogenase enzyme system.

### 5.2 Phosphorus (P) and potassium (K) uptake

#### *5.2.1 Phosphorus and potassium are the most critical soil-supplied macronutrients for soybean after N*

(a) *Phosphorus (P)*: P is essential for ATP synthesis, root development, and seed maturation; soybeans take up inorganic phosphate ($H_2PO_4^-$ and $HPO_4^{2-}$). Because P is relatively immobile, plants explore the soil or employ strategies such as creating a shallow root system with denser hairs, exuding organic acids (e.g., citrate) to solubilize fixed P, and remobilizing P from vegetative tissues to seeds yielding a harvest index up to 81% [9].

(b) *Potassium (K)*: Soybean seeds contain ample K, so the crop removes large amounts during harvest. K activates enzymes, regulates water balance to enhance drought tolerance, and facilitates carbohydrate transport. Rapid uptake occurs mainly between the late vegetative and early reproductive stages (V7 to R2/R3), with up to 90% of total K taken up by R5. Roots absorb $K^+$ primarily through mass flow and diffusion toward the root surface [9].

### 5.3 Micronutrients and environmental influence

Soybeans also require several micronutrients, the availability of which is heavily influenced by soil conditions (**Tables 1** and **2**).

(a) Uptake mechanisms: Most micronutrients are absorbed by the roots *via* diffusion and mass flow, often through specific transporter proteins on the root cell membranes [10].

(b) pH effect: Maintaining the optimal soil pH range (6.0 to 6.8) is critical because it directly controls the chemical form and solubility and thus the availability of most essential mineral nutrients [9].

| Nutrient | Primary role and uptake timing | Availability constraint (Soil pH) |
|---|---|---|
| Iron (Fe) | Chlorophyll production, photosynthesis. | Low availability in alkaline/calcareous soils (high pH), leading to Iron Deficiency Chlorosis (IDC) [8]. |
| Manganese (Mn) | Photosynthesis, enzyme activation. | Low availability in alkaline/high pH soils, often exacerbated by no-till practices [8]. |
| Zinc (Zn) | Seed filling, enzyme function, phytin formation. | Low availability in alkaline/high pH soils. Nearly 40% of total accumulation occurs during seed fill [9]. |
| Boron(B) | Pollen development, flower initiation. | Low availability in sandy soils or during drought. Does not remobilize well from old to new tissue. |
| Molybdenum (Mo) | Essential for the nitrogen fixation enzyme (nitrogenase). | Low availability in acidic soils (low pH) (opposite of other micronutrients) [10]. |

**Table 1.**
*Primary role and uptake timing of different micronutrient.*

| Soil factor | Effect on nutrient availability | Description |
|---|---|---|
| Soil moisture | Vital for nutrient movement | Water dissolves nutrients and transports them to plant roots, facilitating uptake [10]. |
| Soil pH | Influences nutrient solubility and availability | Soil acidity or alkalinity affects availability of nutrients like iron, phosphorus, and others [9]. |
| Organic matter | Enhances soil health and nutrient availability | Addition of organic matter, manures, and compost boosts nutrient availability and microbial activity [11]. |
| Fertilizer application | Provides essential nutrients in correct amounts and timing | Timely, balanced fertilizer supports nutrient supply; for example, early nitrogen enhances seedling growth [10]. |
| Nutrient balance | Balances interactions and availability | Proper nutrient balance and biological factors (e.g., mycorrhizae) optimize nutrient uptake and reduce toxicity [12]. |

**Table 2.**
*Key soil factors affecting nutrient availability in soybean crops.*

## 6. Factors affecting nutrient availability

The availability and uptake of essential nutrients in soybean are governed by three interrelated factors: the concentration of nutrients in the soil solution, the soil's chemical properties, and the intricate interactions among different nutrient elements [9].

### 6.1 Soil solution content

The soil solution an ultrathin film of water that coats each soil particle—acts as the plant root's immediate nutrient reservoir.

(a) *Mass flow*

Mobile nutrients such as nitrate-nitrogen ($NO_3^-$), sulfate ($SO_4^{2-}$), and calcium ($Ca^{2+}$) are carried directly to the root by transpiration. When the water contains higher concentrations of these nutrients and the plant extracts water more quickly, mass flow becomes more efficient [10].

(b) *Diffusion*

Less mobile nutrients, including potassium ($K^+$) and phosphate ($H_2PO_4^-$), reach the root through diffusion, moving from an area of higher concentration in the bulk soil to a lower concentration near the root surface. This movement relies heavily on the soil's water content and the concentration gradient [12].

### 6.2 Root interception

The root system physically invades soil volumes, encountering nutrients directly. This mechanism is particularly vital for immobile nutrients such as phosphorus (P).

### 6.3 Soil chemical properties

The inherent chemistry of the soil controls how nutrients remain available over time.

(a) *Soil pH*

Soil pH indicates how acidic or alkaline the soil is, and it is the primary determinant of nutrient availability in soybean production (**Table 3**).

(b) *Cation exchange capacity (CEC)*

- CEC indicates how well a soil can retain positive nutrients (cations) such as $K^+$, $Ca^{2+}$, $Mg^{2+}$, and $NH_4^+$.
- Soils with a high CEC—typically those rich in clay or organic matter—hold and release these nutrients more efficiently.

| Soil pH level | Effects on nutrient availability | Specific nutrients affected | Consequences |
|---|---|---|---|
| Low pH (Acidic, < 6.0) | Limits availability of macronutrients; suppresses beneficial microbial activity | N, P, K, Ca, Mg, Mo | Phosphorus fixed by Fe and Al; Rhizobium activity reduced; Mo less available [5]. |
| High pH (Alkaline, > 7.0) | Reduces the availability of many micronutrients due to formation of insoluble compounds | Fe, Mn, Zn | Micronutrient deficiencies such as iron-deficiency chlorosis; nutrients form insoluble compounds with Ca [8]. |

**Table 3.**
*Effect of nutrient availability on different pH level.*

- In contrast, soils with a low CEC, like sandy soils, lose nutrients more quickly *via* leaching and usually need more frequent fertilizer applications [12].

(c) *Organic matter (OM)*

- When organic matter decomposes, it liberates nutrients—especially nitrogen (N), phosphorus (P), and sulfur (S)—through mineralization.
- It also acts as a natural chelating agent, binding micronutrients such as Fe, Zn, and Mn, and keeping them soluble for plant use [11].

### 6.4 Interaction between nutrients

Nutrients rarely act alone; the level of one element can influence how the plant absorbs another element (**Table 4**).

### 6.5 Methods of nutrient absorption in soybean crops

Soybean crops primarily take up nutrients through active and passive transport across their root cells, using these processes to absorb macronutrients such as nitrogen, phosphorus, and potassium as well as micronutrients from the soil. This uptake is significantly influenced by symbiotic relationships, especially nitrogen fixation by rhizobia in root nodules and fungal associations such as mycorrhizae, which expand the reach of the root system.

(a) *Mechanisms of absorption*

Passive absorption—This mechanism does not require metabolic energy from the plant; nutrients simply move down their concentration gradient [10].

Active transport—The plant uses metabolic energy to move nutrients against their concentration gradient, carrying them from soil into root cells [9].

Facilitated diffusion—Nutrients shift down their concentration gradient with the help of carrier proteins embedded in the cell membrane [10].

| Type | Nutrient pair | Description |
|---|---|---|
| Antagonism (inhibition) | Potassium (K) vs. magnesium (Mg) vs. calcium (Ca) | High soil K reduces uptake of Mg and Ca as these cations compete for root-binding sites, and vice versa [9]. |
| | phosphorus (P) vs. zinc (Zn) | Excess P interferes with Zn utilization or promotes its precipitation, causing Zn deficiency in soybean [6]. |
| | Manganese (Mn) vs. iron (Fe) | Elevated Fe levels hinder Mn uptake and vice versa, limiting absorption of each other [8]. |
| Synergism (promotion) | Molybdenum (Mo) and biological nitrogen fixation (N) | Adequate Mo is essential for nitrogenase enzyme activity in Rhizobium, enhancing nitrogen fixation efficiency [10]. |
| | Sulfate (S) and nitrogen (N) | Sulfate is vital for amino acid and protein synthesis, supporting effective utilization of nitrogen for high protein seed production [12]. |

**Table 4.**
*Interaction between nutrient in soybean cultivation.*

(b) Symbiotic relationships

*Rhizobia*—Soybean plants form a symbiotic partnership with *Bradyrhizobium* bacteria in their root nodules, fixing atmospheric nitrogen into a usable for an ammonia [11].

Mycorrhizal fungi—These fungi establish a mutualistic relationship with soybean roots, extending the root system and enhancing the uptake of nutrients such as phosphorus and zinc [11].

### 6.6 Fertilizer application techniques influencing nutrient uptake

Fertilizer application techniques (**Table 5**) influence how soybeans absorb nutrients by changing the timing, placement, and form of those nutrients. Key techniques include banding for phosphorus (P), which is especially important early in the season, and foliar feeding for nitrogen (N) to complement soil-applied N without disrupting nitrogen fixation. Integrated nutrient management (NM), which combines organic sources such as vermicompost or farmyard manure with inorganic fertilizers and the biofertilizer Rhizobium, has been shown to significantly raise overall N, P, and K uptake and to improve yield [13].

### 6.7 Nutrient management strategies

Integrated nutrient management (INM)—mixing inorganic fertilizers with organic materials such as vermicompost or farmyard manure (FYM) is highly effective. INM improves the overall uptake of nitrogen, phosphorus, and potassium, increases protein and oil content in the seeds, and enhances soil health [14].

Starter fertilizer—Applying a small amount of nitrogen (20–40 kg $ha^{-1}$) as a starter dose at planting supplies nitrogen during the seedling stage, preventing "nitrogen hunger" without inhibiting nodulation and nitrogen fixation [8].

Balanced fertilization—A balanced mix of N, P, K, and sulfur (S) is crucial for optimal yield; rates should be based on soil tests to avoid unnecessary cost and environmental loss.

| Application method | Description | Key benefits |
|---|---|---|
| Banding | Applying phosphorus fertilizer in a narrow band near the seed | Promotes early root development and energy storage in young soybean plants [10]. |
| Broadcasting | Spreading granular fertilizers evenly across the field | Common for N, P, K but less efficient for phosphorus; may cause nutrient loss [8]. |
| Foliar feeding | Spraying nitrogen fertilizer onto foliage | Meets nitrogen demand during pod-filling, boosts yield without reducing biological nitrogen fixation [12]. |
| Side-dressing | Applying additional fertilizer beside plant rows after establishment | Supplies nutrients during rapid growth phases, especially nitrogen [12]. |
| Soil injection | Injecting liquid or granular fertilizers directly into soil | Improves nutrient placement, reduces losses compared to broadcasting [11]. |

**Table 5.**
*Nutrients application methods in soybean cultivation.*

Rhizobium inoculation—Inoculating seeds with Bradyrhizobium bacteria is essential for biological nitrogen fixation in soybeans and should be done together with other fertilizer applications for maximum benefit [15].

*Other important factors*

- *Soil testing:* Using soil test data to guide fertilizer recommendations ensures that the exact amount of nutrients is applied to meet the crop needs.
- *Organic matter:* Adding organic matter improves soil structure, increases nutrient availability, and stimulates microbial activity, thereby boosting plant nutrient uptake.

## 6.8 Roles and functions of vital nutrients

Description of each essential nutrient's role in crop growth and development is shown in **Table 6**.

| Nutrient | Recommended (based on soil test) | Interpretations |
|---|---|---|
| Nitrogen | 15–25 kg $ha^{-1}$ | Most N is supplied through symbiotic fixation by *Rhizobium* bacteria, so only a starter dose is typically needed. High rates can inhibit nodulation [10]. |
| Phosphorus | 20–80 kg $ha^{-1}$ ($P_2O_5$) | Application rates depend heavily on soil P levels. In low P soils, rates can be higher (e.g., 60–100 kg $ha^{-1}P_2O_5$) [8]. |
| Potassium | 20–50 Kg $ha^{-1}$($K_2O$) | Rates are variable based on the amount of K in the soil, ranging from 40 to 120 kg $ha^{-1}K_2O$ in some regions [12]. |
| Sulfur | 20–40 kg $ha^{-1}$ | Often applied as gypsum (around 220 kg $ha^{-1}$of gypsum) or with PK fertilization [13]. |
| Calcium | 15–20 kg $ha^{-1}$ | Usually supplied by liming the soil to maintain an optimal pH of 6.0–7.0 [6]. |
| Magnesium | 5–10 kg $ha^{-1}$ | Typically supplied *via* liming with dolomite, deficiencies are rare but occur in coarse soils [10]. |

**Table 6.**
*Interpretation and recommended dose of different macronutrient.*

| Nutrient | Deficiencies |
|---|---|
| Nitrogen | General yellowing of the entire leaf, followed by necrosis |
| Phosphorus | Dark to bluish-green leaves with interveinal chlorosis at the base of the leaf. |
| Potassium | Yellowing along the leaf margin, progressing to brown spots and necrosis. Veins may remain green. |
| Sulfur | Yellowing of the entire leaf. |
| Magnesium | Interveinal chlorosis on older leaves. |
| Iron | Interveinal chlorosis with veins remaining green |
| Manganese | Interveinal chlorosis with small necrotic brown spots and eventual necrosis [12]. |
| Zinc | Interveinal chlorosis accompanied by stunted growth |
| Calcium | Premature senescence of younger leaves with small brown spots; collapse of petioles in severe cases. |
| Boron | Stunted growth and swollen nodes; death of growing points. |

**Table 7.**
*Deficiency symptoms of different nutrient.*

### 6.9 Deficiency symptoms and diagnosis

Visual indicators of nutrient deficiencies in soybeans include interveinal chlorosis (yellowing between the veins), necrosis (tissue death), and stunted growth. The location of these symptoms can help identify which nutrient is lacking: deficiencies of mobile nutrients such as nitrogen (N), phosphorus (P), and potassium (K) mainly appear on older, lower leaves, whereas deficiencies of immobile nutrients like iron (Fe), manganese (Mn), and sulfur (S) are seen on newer, upper leaves (**Table 7**) [13].

## 7. Diagnostic approaches in field and laboratory for deficiency confirmation

Confirming nutrient deficiencies in soybean requires a mix of field-based visual diagnosis and laboratory analyses (soil and plant tissue testing); the most reliable approach is to combine all three methods [14].

### 7.1 Field diagnosis

Field diagnosis relies mainly on visual symptoms and is a helpful starting point, but it has limitations because those symptoms may stem from disease, insects, or environmental stress (**Table 8**) [13].

### 7.2 Laboratory diagnosis

Laboratory testing provides clear, quantitative confirmation of nutrient levels and is the most precise way to diagnose deficiencies, especially in hidden-hunger situations where crop yield drops before any obvious symptoms appear (**Table 9**) [11].

| Aspect | Details |
|---|---|
| Symptom Location | Mobile nutrients (N, P, K, Mg) show deficiency symptoms first on older, lower leaves due to reallocation to new growth.<br>Immobile nutrients (S, Ca, Fe, Mn, Cu, Zn) show symptoms first on younger, upper leaves [12]. |
| Comparison | Assess "bad" (poor) versus "good" (healthy) areas in the field to highlight contrasting nutrient conditions [11] |
| Root Examination | For N deficiency, dig up plants and inspect root nodules. Healthy nodules have pink/red interiors; dull or inactive nodules suggest N deficiency (user input). |
| Precision agriculture Tools | Hyperspectral imaging, multispectral images from drones and satellites, and AI-driven smartphone apps provide quick, non-destructive diagnosis of nutrient deficiencies [14].<br>Tools include SPAD meter, GreenSeeker, leaf color charts, multispectral imaging, drones, soil sensors, smartphone apps with AI advisory, automatic GPS-steered tractors, and decision support software [13].<br>These technologies allow site-specific nutrient management by detecting spatial variability of nutrient status within fields for targeted fertilizer application, improving efficiency and reducing waste. |

**Table 8.**
*Visual inspection and precision agriculture tools for identifying nutrient deficiencies.*

| Step | Details |
|---|---|
| Sampling | Collect 15–20 soil core samples at rooting depth (0-8 inches) in a zig-zag pattern parts of the field.<br>Keep samples from different management zones separate to capture variability.<br>Avoid non-representative areas such as fertilizer bands, headlands, or eroded spots [16].<br>Use clean stainless steel probes or augers and clean plastic buckets for composite sampling. |
| Analysis | Send samples to a lab that measures nutrient concentrations such as N, P, K, pH, organic matter, and other soil properties.<br>Lab tests may include pH and micronutrient assessments to detect issues like high-pH induced iron chlorosis [13]. |
| Interpretation | Compare lab results with critical nutrient sufficiency ranges and limits specifically for soybean.<br>Use data primarily for pre-plant fertilization planning and nutrient management decisions [14].<br>Regular testing every 3–4 years recommended; annual testing best for precision but may be costly. |

**Table 9.**
*Soil testing for nutrient evaluation in soybean fields.*

### 7.3 Plant tissue testing

- *Sampling:* Gather the specified plant parts (e.g., the uppermost fully expanded trifoliate leaf) at the desired growth stage usually R2 or full bloom while following the laboratory's sampling protocol, such as whether to include the petiole [13].

- *Analysis:* The lab examines the tissue to determine the exact nutrient concentrations present.

- *Interpretation:* The resulting values are compared with established critical tissue concentrations or sufficiency ranges to confirm a deficiency; tissue testing is useful for diagnosing in-season conditions and for validating soil test results [14].

For the better diagnosis of deficiency symptoms, it is good to combine all the methods, field or visual observations, soil test, and plant tissue analysis.

### 7.4 Sustainable nutrient management strategies

Sustainable nutrient management for soybeans blends Integrated Nutrient Management (INM), which mixes organic and inorganic fertilizers with biological agents and strategic agronomic practices such as crop rotation and carefully timed applications. Key methods are performing soil tests to inform fertilizer decisions, adding organic matter such as compost or manure to enhance soil health, applying nutrients strategically at sowing and throughout later growth stages, and inoculating seeds with biofertilizer [17].

#### *7.4.1 Integrated nutrient management (INM)*

- *Combine organic and inorganic fertilizers:* Using compost or farmyard manure together with synthetic fertilizers boosts soil fertility and improves the efficiency of nutrient use [11].

- *Incorporate biofertilizers:* Apply biofertilizers to stimulate atmospheric nitrogen fixation and to solubilize soil-bound phosphorus [16].

- *Use soil testing:* Prior to fertilization, analyze the soil to determine its exact phosphorus and potassium needs and adjust application rates to avoid over-application [18]. There are many fertilizers application strategy shown in **Table 10**.

### 7.5 Principles of sustainable production in soybeans

(a) *Soil health management:* Sustainability starts with methods that keep the soil undisturbed, such as conservation or no-till farming, to cut erosion and improve water flow. Keeping crop residues on the field and adding organic materials like farmyard manure or compost raises soil organic matter and encourages helpful microbes [22].

(b) *Crop rotation and intercropping:* Switching soybeans with cereals such as maize or wheat breaks pest and disease cycles and boosts soil fertility. Intercropping improves land use and helps control weeds [23].

(c) *Integrated pest management (IPM):* Instead of using only chemicals, IPM blends cultural tactics (e.g., choosing right planting dates and resistant varieties), biological controls (e.g., beneficial insects), physical methods, and chemicals to keep pests and diseases below economic thresholds [24].

(d) *Efficient water management:* Providing water that matches the crop's needs—especially during key periods like flowering and pod filling—is vital. Techniques such as drip irrigation, mulching, and rainwater harvesting conserve water and make its use more effective [25].

(e) *Biodiversity conservation:* Increasing plant diversity inside fields and protecting natural areas like riparian buffers supports pollinators and natural pest enemies, strengthening overall ecosystem resilience [26].

(f) *Variety selection:* Picking soybean varieties that are resilient to the local climate and resistant to pests, diseases, drought, and heat is a key adaptation strategy [27].

| Fertilizer practice | Details |
|---|---|
| Apply phosphorus (P) and potassium (K) at sowing | Apply full recommended amounts of P and K at planting to support early root development and growth [19].<br>Applying P and K ahead of soybean or through previous crop can also be effective based on soil test and crop rotation [20]. |
| Split nitrogen (N) application | Apply part of the nitrogen at sowing and the remainder 25–30 days later to support vegetative and reproductive stages [21].<br>Split N helps meet crop demands and reduce loss. |
| Starter dose of nitrogen at planting | Use a small, low-concentration starter dose of N at planting to prevent early seedling deficiency.<br>Starter N enhances nodulation and biological N fixation [21].<br>Placement of N below nodulation zone (e.g., 8 inches deep) minimizes negative effects on N fixation. |

**Table 10.**
*Strategy of fertilizer application timing and methods for soybean crop.*

Improving soil health in soybean farming means merging organic and mineral fertilization with practices such as inoculating seeds with *Brady Rhizobium japonicum*; this strategy maximizes nutrient availability, enhances soil structure, and lifts yields and quality including protein and oil content while cutting the need for chemical fertilizers [28].

## 8. Conclusion

Effective nutrient management is critical for maximizing soybean crop growth, development, and yield. Soybean plants require a balanced supply of 15 essential nutrients, including six macronutrients and nine micronutrients, each with specific roles and concentrations necessary for optimal physiological functions. Deficiencies or excessive concentrations of particular nutrients can adversely affect nutrient uptake, plant health, and productivity. Balanced nutrient supply ensures that soybeans can express their full genetic potential, cope with biotic and abiotic stresses, and achieve sustainable production. Understanding nutrient mobility within the plant, availability forms in soil, nutrient interactions, and appropriate fertilizer formulations is essential to designing effective nutrient management strategies. Moreover, visual diagnosis of nutrient deficiencies through observable symptoms enables timely and cost-effective nutrient interventions in the field, even when soil and tissue analyses are not readily available. Integrating these approaches enhances nutrient use efficiency, supporting healthy growth and high yields while minimizing environmental impacts. Thus, comprehensive and precise nutrient management underpins sustainable soybean production systems, ensuring economic viability and resource conservation.

## Acknowledgements

The author acknowledges the use of AI (Trinka) for language polishing of the manuscript.

## Authors' contributions

All authors have equal contributions.

## Conflicts of interest

The author declares no conflicts of interest.

## Author details

Umesh Kumar Singh[1*], Santosh Korav[1], Dhanpreet Kaur[1],
Thangavel Palanivel Swarnam[2] and Niharika Kodirekka[3]

1 Department of Agronomy, Lovely Professional University, Punjab, India

2 ICAR - Indian Institute of Farming Systems Research, Modipuram Meerut, Uttar Pradesh, India

3 Department of Soil Science, Sardar Vallabhbhai Patel University of Agriculture and Technology, Meerut, Uttar Pradesh, India

*Address all correspondence to: 1999umesh8@gmail.com

## References

[1] Usman T, Ramzan T. Agronomic Practices for Soybean Production under Climate-Resilient Conditions. 2nd ed. London, UK: IntechOpen; 2002. pp. 1-13. DOI: 10.5772/intechopen.1012006

[2] Bagale S. Nutrient management for soybean crops. International Journal of Agronomy. 2021;**2021**(1):304-634. DOI: 10.1155/2021/3304634

[3] Van K, Kim MY, Shin JH, Kim KD, Lee YH, Lee SH. Molecular evidence for soybean domestication. Genomics of plant genetic resources. In: Managing, Sequencing and Mining Genetic Resources. Vol. 1. Netherlands: Springer; 2013. pp. 465-481. DOI: 10.1007/978-94-007-7572-5_19

[4] USDAF. World Agricultural Production. Washington, D.C: Bange, G., USDA World Agricultural Outlook Board; 2022. pp. 09-10

[5] Vishnoi NK. An overview on soybean. Asian Journal of Multidimensional Research. 2021;**10**:205-210

[6] McCauley A, Jones C, Jacobsen. Plant nutrient functions and deficiency and toxicity symptoms. Nutrient Management Module. 2009;**9**:1-16

[7] Arnarson A. Soybeans Nutrition Facts and Health Effects. Healthline; 2019. Available from: https://www.healthline.com/nutrition/foods/soybeans [Accessed: March 29, 2020]

[8] Ohyama T, Tewari K, Ishikawa S, Tanaka K, Kamiyama S, Ono Y, et al. Role of nitrogen on growth and seed yield of soybean and a new fertilization technique to promote nitrogen fixation and seed yield. In: Soybean: The Basis of Yield, Biomass and Productivity. Hoboken, New Jersey, USA: Wiley; 2017. pp. 153-218

[9] Hungria M, Nogueira MA, Araujo RS. Co-inoculation of soybeans and common beans with *rhizobia* and *azospirilla*: Strategies to improve sustainability. Biology and Fertility of Soils. 2013;**49**:791-801

[10] Dabesa A, Tana T. Response of soybean (*Glycine max* L.) to Bradyrhizobium inoculation, lime, and phosphorus applications at Bako, Western Ethiopia. International Journal of Agronomy: Wiley. 2022;**1**:6686957. DOI: 10.1155/2021/6686957

[11] Hellal FA, Abdelhamid MT. Nutrient management practices for enhancing soybean (*Glycine max* L.) production. Acta Biologica Colombiana. 2013;**18**:239-250

[12] Pedrozo A, Oliveira NJ, Alberton O. Biological nitrogen fixation and agronomic features of soybean (*Glycine max* Merr.) crop under different doses of inoculant. Acta Agronomica. 2018;**67**:297-302

[13] Abuli JS, Mugwe JN, Mugendi DN. Effects of phosphorus sources on soybean yield in central highlands of Kenya. Scientific Conference Proceedings. 2013;**13**:595-601

[14] Enderson JT, Mallarino AP, Haq MU. Soybean yield response to foliar-applied micronutrients and relationships among soil and tissue tests. Agronomy Journal. 2015;**107**:2143-2161

[15] De L, Lopes J, Rambo LO, Marafon TP, Oliveira S, Souza L, et al. Soybean yield responses to micronutrient fertilizers. In: Soybean-the Basis of Yield, Biomass and Productivity. London, UK: IntechOpen; 2017. DOI: 10.5772/67157

[16] Rodrigues VA, Crucial CA, Bossolani JW, Moretti LG, Portugal JR, Mundt TT, et al. Magnesium foliar supplementation increases grain yield of soybean and maize by improving photosynthetic carbon metabolism and antioxidant metabolism. Plants. 2021;**10**:797-811

[17] Ma J, Song Z, Yang J, Wang Y, Han H. Cobalt ferrite nanozyme for efficient symbiotic nitrogen fixation via regulating reactive oxygen metabolism. Environmental Science: Nano. 2021;**8**:188-203

[18] Barth G, Francisco E, Suyama JT, Garcia F. Nutrient uptake illustrated for modern, high-yielding soybean. Better Crop Plant Food. 2018;**102**:11-14

[19] Jat G, Sharma SK, Meena RH, Choudhary R, Choudhary RS, Yadav SK. Studies on effect of zinc application on quality and yield of soybean (*Glycine max* L.) under typic haplustepts soil. Indian Journal of Pure and Applied Biosciences. 2021;**9**:188-193

[20] Lavres J, Castro FG, Sousa GM. Soybean seed treatment with nickel improves biological nitrogen fixation and urease activity. Frontiers in Environmental Science. 2018;**4**:37. DOI: 10.3389/fenvs.2016.00037

[21] Banerjee P, Das P, Sinha S. Importance of molybdenum for the production of pulse crops in India. Journal of Plant Nutrition. 2021;**45**:300-310. DOI: 10.1080/01904167.2021.1952226

[22] Rout GR, Sahoo S. Role of iron in plant growth and metabolism. Reviews in Agricultural Science. 2015;**3**:1-24. DOI: 10.7831/ras.3.1

[23] Pagano MC, Miransari M. The importance of soybean production worldwide. In: Abiotic and Biotic Stresses in Soybean Production. Cambridge, Massachusetts, USA: Academic Press; 2016. pp. 1-26. DOI: 10.1016/B978-0-12-801536-0.00001-3

[24] KS S, Dharmaraj BM. System of soybean intensification: New method of soybean cultivation. Environment;**659**:200-210

[25] Hamza M, Basit AW, Shehzadi I, Tufail U, Hassan A, Hussain T, et al. Global impact of soybean production: A review. Asian Journal of Biochemistry, Genetics and Molecular Biology. 2024;**16**:12-20. DOI: 10.9734/AJBGMB/2024/v16i2357

[26] Khojely DM, Ibrahim SE, Sapey E, Han T. History, current status, and prospects of soybean production and research in sub-Saharan Africa. The Crop Journal. 2018;**6**:226-235

[27] Takamoto A, Togami K, Uchida T, Maekawa T, Takahashi T, Namikawa M, et al. Exploring the impact of shoot nutrient uptake on soybean yields in a paddy-upland rotation field in Akita, Japan: A path analysis approach. Soil Science and Plant Nutrition. 2025;**71**:574-580. DOI: 10.1080/00380768.2025.2503325

[28] Faé GS, Kemanian AR, Roth GW, White C, Watson JE. Soybean yield in relation to environmental and soil properties. European Journal of Agronomy. 2020;**118**:1161-0301. DOI: 10.1016/j.eja.2020.126070

Chapter 3

# Competitive Dynamics and Yield Performance of Soybean-Based Intercropping Systems on Inceptisols

*Kristamtini, Arini Putri Hanifa, Setyorini Widyayanti, Arlyna Budi Pustika, Sutardi, M. Basir Nappu, Susilawati, Evert Y. Hosang, Muhammad Iskandar Ishaq, I. Gusti Komang Dana Arsana, Yustisia, Sugeng Widodo, Forita Dyah Arianti, Heni Purwaningsih and Atman*

## Abstract

This study evaluated the competitive dynamics and yield performance of soybean-based intercropping systems on dryland Inceptisols. Conducted in Gunungkidul, Yogyakarta, Indonesia, from March to August 2020, the research compared four intercropping systems: upland rice-soybean, maize-soybean, maize-peanut, and maize-upland rice, against their respective monocultures. A randomized complete block design with five replicates was employed. Analysis to assess agronomic performance includes Land Equivalent Ratio (LER), Aggressivity (A), Competitive Ratio (CR), Relative Crowding Coefficient (K), and Actual Yield Loss (AYL). Results clearly demonstrated that maize-soybean intercropping was the most advantageous system for Inceptisols. While maize exhibited greater competitive dominance over soybean, as evidenced by Land Equivalent Ratio (LER) of 1.91, Actual Yield Loss (AYL) of 1.07, Aggressivity (A), and Competitive Ratio (CR), the intercropping system significantly enhanced overall productivity. Notably, maize-soybean intercropping achieved the highest yields. This study highlights the superior competitiveness and productivity of soybean when intercropped with maize on dryland Inceptisols, providing valuable insights for optimizing agricultural practices in these challenging environments.

**Keywords:** dryland, inceptisol, competitive value, sustainable agriculture, yield advantage

## 1. Introduction

Many studies have been undertaken around the world to determine the best management strategies for dryland agriculture and farming. Roughly one-third of the world's population makes their livelihoods from drylands, which make up 40% of all cultivated land [1]. Global warming and climate change affect the future productivity of dryland agriculture as they bring profound consequences of increasing the severity, frequency, and duration of floods and/or droughts. These consequences, in turn, result in high or low runoff rates, more water shortages or abundances, decreased water productivity, and diminished crop yields [2]. Eliminating yield gaps in dryland agriculture and stabilizing crop yields over time are critical to meeting global food needs, especially in response to global population growth and altering weather patterns. In order to address food security, agricultural intensification has long been thought to be the most effective strategy [3]. However, because of its negative consequences on biodiversity, overreliance on chemical inputs, and agricultural production systems [4, 5], the sustainability of intensive agriculture is questioned.

The transition from the current near-monocropping to a diversified cropping system is acknowledged as a potential agricultural technique to improve agricultural sustainability and food security [6]. Intercropping, a time-tested practice, grows different crops together to create a more diverse ecosystem. This not only reduces the need for harmful chemicals but also improves the overall health of the farming environment [7].

Intercropping has been named the "new green revolution" since it can improve farming productivity by utilizing data on how species complement one another and provide a way to increase rural's economic level [8]. The intercropping framework contributes to 15–20% of the food supply to the world [9] and promotes farming yield per unit of land [10]. Furthermore, several studies have reported that an intercropping system can promote soil fertility if the plants are intercropped with legumes [11] and potentially reduce the application of N fertilizer [12]. Numerous research on intercropping in dryland environments have shown its positive effect on crop productivity. Despite intercropping's benefits, most intercropping farmers continue to obtain low yields due to unregulated spacing and incompatible plant combinations. The development of an efficient intercropping system is essential for realizing increased system productivity and profitability. This is particularly important for resource-poor farmers in developing countries, where such a system could significantly improve their livelihoods [13].

In the developing country of Indonesia, the dryland area reaches 76.20% of the land area or 144.47 million ha, spread across Sumatra, Kalimantan, Sulawesi, Papua, Java, Maluku, and Bali and Nusa Tenggara, with areas of 33.25, 41.61, 16.57, 28.60, 10.27, 7.45, and 6.70 million ha, respectively. Dryland in Indonesia is categorized into three soil orders, namely, Ultisols, Oxisols, and Inceptisols [14]. Representing a 37.5% of mainland Indonesia [15], Inceptisols cover an area of 52 million ha [16]. Although Inceptisols encompass a large proportion of the dryland area and have the potential for productive farming, the utilization of Inceptisols for farming is still hindered by several challenges and issues. Inceptisols are underutilized due to a lack of nutrients and low organic matter, which can demotivate farmers in utilizing such kind of soil.

Farmers' motivation in implementing intercropping is ultimately determined by how profitable the system is [17]. Hence, financial advantages become an essential aspect that needs to be considered when deciding whether the farmer should implement the intercropping system. The intercropping system uses relatively fewer energy inputs, such as pesticides and fertilizers, which leads to less cost. This offers a significant pathway for decreasing greenhouse gas emissions that are directly and indirectly derived from agricultural resource inputs [18].

Lower fertilizer input in the intercropping system is contributed by legumes in the intercropping combination. Legumes have a role in atmospheric nitrogen fixation and release some fixed nitrogen into the soil or become available for the companion crop (usually cereal) through root decomposition. Cereal+legume intercropping enhances soil health and system productivity while offering more opportunities to reduce the negative effects of moisture and nutrient stress [19]. Maize-cowpea intercropping not only boosted forage productivity and nutrient uptake but also helped farms achieve protein self-sufficiency by increasing on-farm protein production [20]. However, making sure that component crops do not compete with one another for sunlight, moisture, nutrients, and space is crucial. Soybean (*Glycine max* (L). Merr), peanut (*Arachis hypogaea*), upland rice (*Oryza sativa*), and maize (*Zea mays*) are among the legumes and cereals grown in the dryland as monoculture and or intercropping. Finding the best combination of them in the dryland inceptisol benefits farmers.

Various indices have been proposed to evaluate the relative advantage of intercropping over monocropping and the competition between cereal and legume intercrops, a combination that enhances soil health and system productivity. These indices include the Land Equivalent Ratio (LER), crop equivalent yield, aggressivity (A), Actual Yield Loss (AYL), and Competitive Ratio (CR) [19]. Another study focuses on the relative crowding coefficient (K), competition ratio, yield loss, growth improvement, intercropping dominance index, and light irradiance [20]. Meanwhile, previous studies focusing on the critical aspects of intercropping to find the appropriate system that would guarantee productivity have been conducted. Examples of these studies focus on breeding variety, screening and planting patterns, fertilizer management, its correct application, and water use efficiency and distribution [21]. However, no specific studies have compared the competitive value and yield advantages of cereal+legume combinations and their profitability, particularly in Inceptisols. Therefore, the objective of this study is to evaluate the competitive value and the yield of various intercropping systems in dryland Inceptisols.

## 2. Methods

### 2.1 Timing, site study, and planting material

The research was conducted in the dry season from March and August 2020 in Inceptisols at Trengguno Kidul, Ponjong, Sidorejo, Gunungkidul, Yogyakarta (7°59'4" N, 110°41'3" E; at about 226.5 m a.s.l), Indonesia. The climate consisted of C2 with 4–5 wet months (December, January, February, and March) and 6–7 dry months (April, June, July, August, September, October, and November). The climate information (precipitation and the number of rainy days) during the study was shown in **Figure 1**. The soil had the following characteristics: pH $H_2O$ (7.41), C-organic (1.33%), N total (0.09%), $P_2O_5$ (56 ppm P), K-dd (0.14 cmol/kg), and C/N ratio (15). Four systems of intercropping: upland rice-soybean (USI), maize-upland rice intercropping (MUI), maize-soybean intercropping (MSI), and maize-peanut intercropping (MPI) were evaluated. Planting materials were obtained from Indonesian food crops research centers. Rice seeds were obtained from the Indonesian Rice Research Institute; soybean and peanut seeds were from the Indonesian Legume and Tuber Crops Research Institute. Maize seeds were obtained from the Indonesian Cereal Crops Research Institute. Five replicates of each intercropping combination were randomly assigned and compared to the productivity of two hectares of the monoculture.

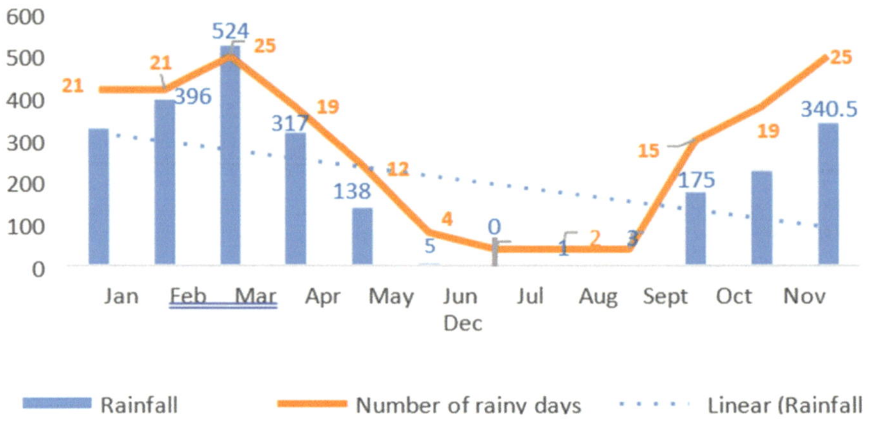


**Figure 1.**
*Month-to-month rainfall and number of rainy days in 2020.*

## 2.2 Four intercropping systems

### *2.2.1 Upland rice-soybean intercropping (USI)*

Using a randomized design block with five replicates, three new high-yielding rice cultivars (Inpago 10, Inpari 42, and Inpari 43) and two soybean cultivars (Dega 1 and Biosoy 2) were employed. Upland rice was sown at 20 cm between columns and 10 cm between rows and planted in four columns, whereas the space between rice and soybean squares was 30 cm. The space between lines of soybeans was 30 cm × 10 cm, planted in four columns (**Figure 2**).

### *2.2.2 Maize-soybean intercropping (MSI)*

The design block in this category was randomized in five replicates by combining three maize cultivars (Srikandi Ungu, Pulut Uri 1, Nasa 29) and two soybean cultivars (Dega 1 and Biosoy 2). The planting distance for the maize was 70 cm × 25 cm, planted in two rows. The distance between maize and soybean blocks was 40 cm, and the planting distance for the soybean was 30 cm × 10 cm (planted in four columns) (**Figure 3**).

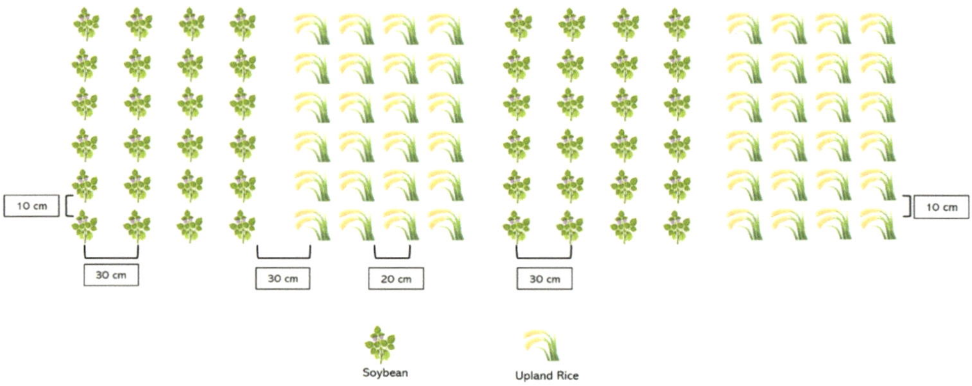


**Figure 2.**
*Layout of upland rice-soybean intercropping.*

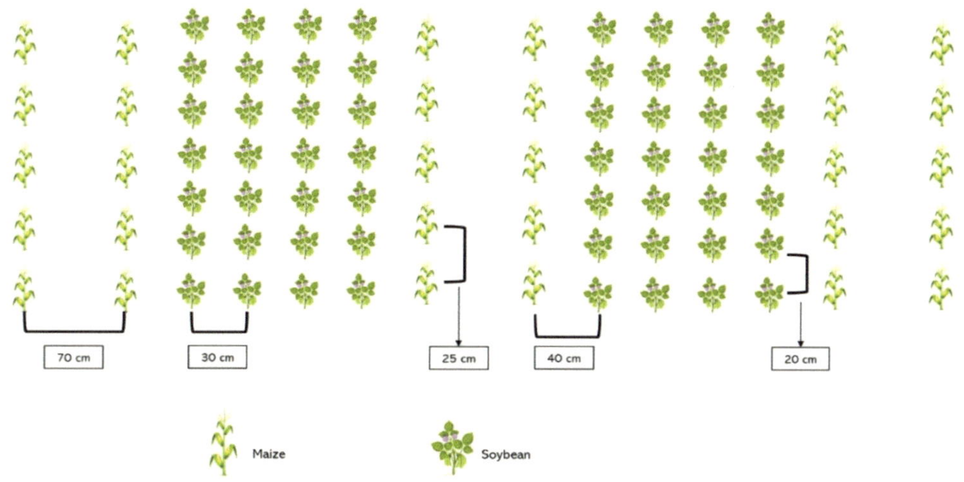


**Figure 3.**
*Layout of maize-soybean intercropping.*

### *2.2.3 Maize-upland rice intercropping (MUI)*

MUI was randomized with five replicates using one hybrid maize (Bisi 18) and three upland rice cultivars (Inpago 10, Inpari 42, Inpari 43). The planting space for maize was 70 cm × 25 cm between columns for the two lines, and the space between maize and rice blocks was 40 cm. The planting space for upland rice was 20 cm × 15 cm, planted in four columns (**Figure 4**).

### *2.2.4 Maize-peanut intercropping (MPI)*

The design block for MPI was randomized with five replicates using (Bisi 18 -a hybrid maize) and two peanut cultivars (Talam 3 and 1 local peanut). The spacing between two rows of maize was 300 cm × 25 cm, and the spacing between maize and

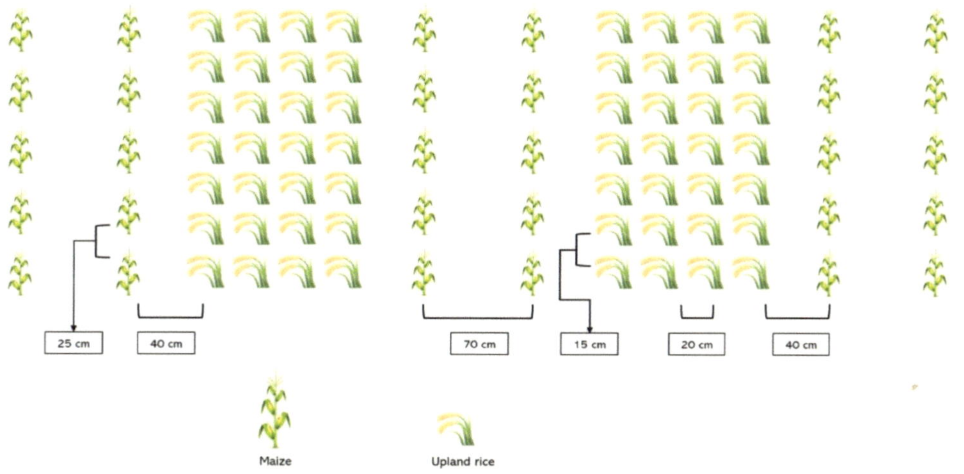


**Figure 4.**
*Layout of maize-upland rice intercropping.*

peanut blocks was 50 cm. The planting space of peanuts was 20 cm × 20 cm, planted in 11 columns (**Figure 5**).

## 2.3 Data collection

Yield data were collected for all intercropping types (MUI, USI, MSI, and MPI). The parameters observed were LER, CR, A, K, and AYL. The data were analyzed using ANOVA, followed by Duncan Multiple Range Test (DMRT) *post hoc* with a significance value of $p = 0.05$.

## 2.4 Data analysis

### *2.4.1 Land equivalent ratio (LER)*

The Land Equivalent Ratio (LER) evaluates how effectively land is utilized by measuring the yield of harvested crops. This ratio helps determine the performance and productivity of an intercropping system. As noted by reference studies [22, 23], the LER can be expressed using the following formula:

$$LER = \frac{Yab}{Yaa} + \frac{Yba}{Ybb} \tag{1}$$

Where:

- Yab represents the yield of crop (a) in the intercropping system,
- Yaa is the yield of crop (a) in the monocropping system,

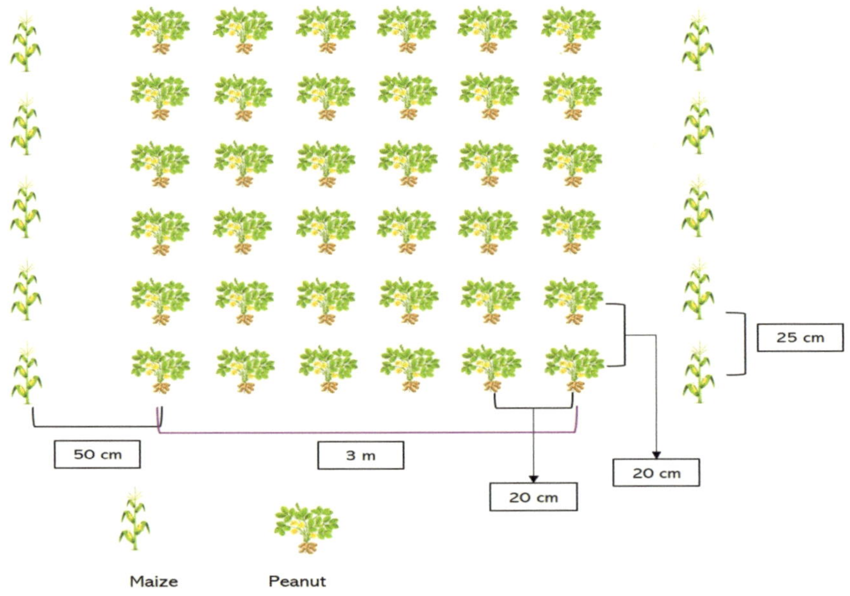


**Figure 5.**
*Layout of maize - peanut intercropping.*

- Yba represents the yield of crop (b) in the intercropping system, and
- Ybb is the yield of crop (b) in the monocropping system.

*2.4.2 Competitive ratio (CR)*

The Competitive Ratio (CR) refers to the LER for each crop species grown in an intercropping system from the start of planting. This metric also accounts for the proportion of each crop in the planting arrangement [11]. The CR, also known as the crowding ratio, is utilized to evaluate the competitiveness between different species. It is calculated using the following formula [24]:

$$CR = \frac{Yab / Yaa}{Yba / Ybb} x \frac{Zba}{Zab} \tag{2}$$

or

$$\frac{LERa}{LERb} x \frac{Zba}{Zab} \tag{3}$$

Where:

- Yab = yield of crop (a) in the intercropping system,
- Yaa = yield of crop (a) in the monocropping system,
- Yba = yield of crop (b) in the intercropping system,
- Ybb = yield of crop (b) in the monocropping system,
- Zba = proportion of crop (b) in the intercropping system, and
- Zab = proportion of crop (a) in the intercropping system.

*2.4.3 Aggressivity (a)*

Intercropping often results in competition between intercropped plant species. The A measures a plant's ability to compete with other species to provide a high yield from an intercropping system. Reference [24] introduced the following formula for calculating A:

$$Aab = \frac{Yab}{Yaa - Zab} - \frac{Yba}{Ybb - Zbb} \tag{4}$$

Where

- Yab = yield of the first crop (a) in intercropping system,
- Yaa = yield of the first crop (a) in monocropping,

- Yba = yield of the second crop (b) in intercropping system,
- Ybb = yield of the second crop (b) in monocropping,
- Zba = proportion of plant b in intercropping,
- Zab = proportion of plant a in intercropping.

*2.4.4 Relative crowding coefficient (K)*

The K measures plant density in intercropping, which denotes the number of plants that can grow in a certain land unit and provide a maximum result [24]. An equation for calculating K as follows [25]:

$$Kab = \frac{Yab \ x \ Zba}{(Yaa - Yab) x \ Zab} \tag{5}$$

Where:

- Yab is = yield of the first crop (a) in intercropping system,
- Yaa = yield of the first crop (a) in monocropping,
- Yba = yield of the second crop (b) in intercropping system,
- Ybb = yield of the second crop (b) in monocropping,
- Zba = proportion of plant b in intercropping,
- Zab = proportion of plant a in intercropping.

*2.4.5 Actual yield loss (AYL)*

The AYL is obtained by comparing the yield loss or gain from intercropped plants to their monoculture crop [11]. The AYL from each species reflects the proportionate yield loss or gain resulting from intercropping relative to their yield in monoculture [26]. The AYL was calculated according references studies [11, 25]:

$$AYL = \left(\frac{Yab}{Xab}\right) / \left(\frac{Yaa}{Xaa}\right)^{-1} \tag{6}$$

Where:

- Yab = yield of the first crop (a) in intercropping system,
- Yaa = yield of the first crop (a) in monocropping,
- Xab = proportion of plant a in intercropping,
- Xaa = proportion of plant a in monocropping.

## 3. Results

### 3.1 Land equivalent ratio (LER)

This study employs the LER model to determine land-use efficiency. The LER determines the effects of competitive and yield advantage from a land perspective while estimating the yield advantage from a time point. The plant growth in the field is presented in **Figure 6**.

MSI had the highest total LER compared to other intercropping packages. The MSI was also more profitable than the single-planting crop. The intercropping pattern between maize and soybeans gave an LER value >1, indicating that this combination is feasible to be applied on dryland Inceptisols in Gunungkidul (**Table 1**).

### 3.2 Competitive ratio (CR)

The CR measures the plant's ability to acquire resources both horizontally and vertically. The results showed that maize had a higher average CR than MSI and MPI, as shown in **Table 2**. The CR values for maize in the two combinations (MSI and MPI) were > 1, while soybean and peanut had CR values of <1. This result indicates

**Figure 6.**
*Visualization of intercropping combinations in dryland Inceptisols. (a) Maize-soybean intercropping, (b) Upland rice-soybean intercropping, (c) Maize-upland rice intercropping, and (d) Maize-peanut intercropping.*

| Intercropping system | Land equivalent ratio | | Total LER |
|---|---|---|---|
| | First crop | Second crop | |
| MSI | 1.55 (maize) | 0.36b (soybean) | 1.91 |
| USI | 0.38 (upland rice) | 0.45ab (soybean) | 0.83 |
| MUI | 1.02 (maize) | 0.29b (upland rice) | 1.30 |
| MPI | 0.47 (maize) | 0.69a (peanut) | 1.17 |

*Note: Values with the same letters are not significantly different according to DMRT at $\alpha$ = 0.05.*

**Table 1.**
*Land equivalent ratio (LER) of intercropping system in dryland Inceptisols.*

| System | CR | | A | | K | | AYL | | ∑ AYL |
|---|---|---|---|---|---|---|---|---|---|
| | 1st crop | 2nd crop | 1st crop | 2nd crop | 1st crop | 2nd crop | 1st crop | 2nd crop | |
| MSI | 2.88a | 0.19b | 5.52a | −5.52b | −13.36a | 0.17b | 1.88a | −0.81b | 1.07a |
| USI | 0.38ab | 0.45b | −1.04b | 1.04a | 0.62a | 1.02ab | −0.65b | −0.51a | −1.16b |
| MUI | 0.55b | 0.53a | −0.08b | 0.08a | 36.25a | 1.47a | −0.45b | −0.47a | −0.92ab |
| MPI | 1.42ab | 0.23b | 0.92b | −0.92a | 8.1a | 1.14ab | 0.42ab | −0.77b | −0.35ab |

*Note: Values with the same letters are not significantly different according to DMRT at $\alpha$ = 0.05.*

**Table 2.**
*Average value of competitive ratio (CR), aggressivity (A), relative crowding coefficient (K), and actual yield loss (AYL).*

that soybean and peanut are suitable for intercropping with maize because maize is taller than soybean and peanut, leading to fair competition of nutrients, space, water, and light.

### 3.3 Aggressivity (A)

The A denotes a plant's dominance over other plant species in intercropping. Maize was found to be more aggressive than soybean and peanut in their respective intercropping combinations (**Table 2**). Maize dominates soybean and peanut in competing for resources vertically and horizontally to enhance production. Aggressive values are distinguished by negative (−) or positive (+) signs.

### 3.4 Relative crowding coefficient (K)

The K values showed that USI, MUI, and MPI were favorable, while MSI was unfavorable. The K for MSI was negative (−) since the soybean plants had a value of $\leq$1 (**Table 2**). When K $\geq$ 1, there is a yield advantage; K = 1, there is no yield advantage; K $\leq$ 1, there is a loss/no profit [26].

### 3.5 Actual yield loss (AYL)

The AYL measures the yield loss or gain of intercropped plants compared to each monoculture crop. The results showed that soybean, rice, and peanut had the highest yield loss in the intercropping, with values for all types of intercropping ranging from −0.47 to −0.81 or equivalent to 47–81%. (**Table 2**). There was no yield loss for maize in the MSI because it had a positive AYL value, as shown in **Table 2**. However, since the maize plant had a positive AYL value and the peanut plant had a negative AYL value, the total AYL value for MPI became negative.

## 4. Discussion

### 4.1 Land equivalent ratio (LER)

Increased maize yields occur due to the abundance of N produced by soybean plants intercropped with maize [11]. Maize roots are intertwined with soybean, allowing them to absorb available nitrate to increase productivity. This increases overall maize productivity and optimizes resource utilization within the farm environment [27]. The increase in maize yield indicates that maize-soybean intercropping (MSI) has the potential to provide significant yields under dryland conditions such as Inceptisols. The presence of soybean as an intercrop not only provides a significant nutritional contribution but also creates a favorable growth environment for maize, resulting in satisfactory yields on land that may have previously been less productive [28]. The intercropping pattern between maize and soybean gave LER values >1, indicating that this combination is feasible to be applied in Inceptisols drylands in Gunungkidul (**Table 1**). These results are in line with reference [29], which reported that the corn-soybean intercropping framework basically has higher land use efficiency and biological prowess (LER > 1). Ceunfin et al. [30] illustrated that when LER = 1, monoculture crops require a larger crop area than intercropping systems to produce comparable yields. The intercropping combinations with LER > 1 are MUI and MPI. This study's findings align with recent research by reference [31], which demonstrated that soybean intercropping with vetiver, as well as the addition of monoculture, resulted in Land Equivalent Ratios (LER) exceeding 1. According to reference [32], intercropping oats and common vetch yielded LER values from 1.03 to 1.09, demonstrating improved land utilization. The research by reference [33] indicated that maize-soybean strip intercropping has a land equivalent ratio greater than 1, showing the positive effects of intercropping while also highlighting maize's stronger competitive position.

Based on the LER (**Table 1**), the most suitable intercrops that are expected to provide high yields in Inceptisols dryland are maize (C4 crop) and soybean (C3 crop), indicating that maize is an excellent partner for soybean. Soybean, as a legume crop, has the ability to fix nitrogen from the air into the soil through bacteria living in its roots [34]. Soybean and maize have complementary traits that allow them to coexist harmoniously. Soybean is a nitrogen-fixing C3 crop, while maize is a nitrogen-consuming C4 crop. Both are in different ecological niches, where soybean is in a lower ecological niche than maize [35]. In addition, maize shading does not affect soybean yield, as soybean is known as a shade-tolerant C3 crop [36]. These conditions create an ideal growth environment for soybeans, which can grow optimally under maize shade without significantly competing for sunlight [37]. These distinctive traits increase the productivity of both maize and soybean [38]. Thus, maize-soybean intercropping can be adopted in any region where both crops are grown. The intercropping combination of C3 and C4 crops can be used for other intercropping systems that have LER > 1, such as MUI and MPI.

Different results were obtained at USI that had LER < 1 (**Table 1**). The low LER occurred because upland rice and soybean are C4 crops, so intercropping these two crops on Inceptisol dryland in Gunungkidul is not profitable.

### 4.2 Competitive ratio (CR)

Competition within intercropping describes the capacity of a species to more effectively exploit limited resources than its intercropped counterpart [39]. Based on CR values, soybean and peanut are suitable for intercropping with maize because maize plants are taller than soybean and peanut, resulting in fair competition for nutrients, space, water, and light. The results of reference [40] showed that crops with CR $\leq$ 1 are mild competitors, suitable for intercropping with crops that have CR $\geq$ 1. Intercropping systems with average CR values <1, namely, MUI and USI, indicate that if these crops are intercropped, they will compete unhealthily for resources and are not suitable for intercropping in Inceptisol drylands in Gunungkidul. Rice and soybean are not suitable for intercropping because these crops are relatively tall, causing intense competition. Upland rice and maize are also not suitable because maize provides shade that is not suitable for rice that is highly dependent on light to increase yield [41]. Maize outperforms soybean and groundnut in vertical and horizontal competition for resources to increase yield. This indicates maize's superiority in vertical and horizontal competition for resources. Plants that perform well in competition will produce the largest yields at the end of their life cycle [30]. The presence of suitable CR in maize and soybean combinations indicates that they can support each other without interfering with each other's growth or yield. Maize provides the right shade structure for soybean without inhibiting its growth, while soybean contributes important nitrogen to maize without interfering with maize growth. This balance in competition for resources results in an optimal growth environment for both crops, which in turn increases the overall productivity of the intercropping system.

### 4.3 Aggressivity (A), relative crowding coefficient (K), and actual yield loss (AYL)

To accurately assess competition in intercropping, actual yield loss (AYL) is used as an efficient index, reflecting the yield performance of individual crops within the mixture [42]. There was no yield loss in maize in MSI because it had a positive total AYL value and a negative total AYL value for MPI. Pea crops with AYL values between −0.42 to −0.02 were equivalent to yield losses of 42–2% compared to monoculture crops [11]. Furthermore, reference [26] also found that yield losses of intercropped cereals and vetches ranged from −0.391 to −0.040, identical to yield losses of 4–39.1%. The overall results of the yield evaluation of competing and dominant crops in the MSI, MUI, and MPI systems showed that maize was stronger and more aggressive in competing for resources, both vertically and horizontally. However, maize had a negative total AYL in the MUI and MPI systems. The plant population on the land unit caused yield loss for maize in the intercropping system with rice and groundnut. In USI, both crops experience yield loss as they compete fiercely for resources. In general, parameters such as LER, CR, K, and AYL indicate that maize-soybean is the most suitable combination for Inceptisols in Gunungkidul dryland.

Such intense competition for resources can result in reduced productivity of maize as it has to compete with other crops for nutrients, water, and sunlight needed for its optimal growth. Although maize is generally proven to be an aggressive crop in competing for resources, its presence in some intercropping systems such as MUI and MPI can lead to suboptimal yields. Intense competition between maize and other crops in terms of resource access can result in an overall decrease in maize yield. This result is in line with the results of research conducted by [43], where intercropping is a sustainable and effective cultivation system for maize and soybean. In this case, while maize has an advantage in competition for resources, in the context of a given intercropping system, strategic adjustments are needed to ensure that its presence does not hamper the yields of other crops in the system [44]. Therefore, it is important to consider the complex interactions between crops in an intercropping system and ensure that a balanced and efficient distribution of resources can be organized to optimally support the productivity of the entire system.

## 5. Conclusion

The most profitable intercropping system in the dryland Inceptisols of Gunungkidul, Yogyakarta is maize-soybean intercropping. Maize plants are more aggressive and competitive than soybean plants, so the intercropping has favorable LER, CR, A, and AYL, which contribute to its profitability. The MSI yielded the highest LER and AYL, with a total of 1.91 and 1.07, respectively. The maize-soybean intercropping can be widely applied to initiate dryland Inceptisols management reform, which can increase food reserves in the respective area.

## Acknowledgements

The authors would like to thank Mr. Edi Januari and Mr. Harjiyo, who have helped with research activities. This research was funded by the Ministry of Agriculture, Republic of Indonesia, through the IAARD program.

## Conflict of interest

The authors declare no conflict of interest.

**Author details**

Kristamtini[1], Arini Putri Hanifa[1*], Setyorini Widyayanti[1], Arlyna Budi Pustika[1], Sutardi[1], M. Basir Nappu[1], Susilawati[1], Evert Y. Hosang[1], Muhammad Iskandar Ishaq[1], I. Gusti Komang Dana Arsana[1], Yustisia[1], Sugeng Widodo[2], Forita Dyah Arianti[3], Heni Purwaningsih[4] and Atman[5]

1 Research Center for Food Crops, National Research and Innovation Agency, Cibinong, Indonesia

2 Research Center for Behavioral and Circular Economics, Research Organization for Governance, Economy, and Community Welfare, South Jakarta, Java, Indonesia

3 Research Center for Sustainable Production System and Life Cycle Assessment, National Research and Innovation Agency, South Tangerang, Indonesia

4 Research Center for Food Technology and Processing, National Research and Innovation Agency, Yogyakarta, Indonesia

5 Research Center for Horticulture, National Research and Innovation Agency, Cibinong, Indonesia

*Address all correspondence to: arin010@brin.go.id

## References

[1] Maestre FT, Quero JL, Gotelli NJ, Escudero A, Ochoa V, Delgado-Baquerizo M, et al. Plant species richness and ecosystem multifunctionality in global drylands. Science. 2012;**335**(6065):214-218

[2] Naorem A. Drylands: An Introduction Enhancing Resilience of Dryland Agriculture under Changing Climate. Singapore: Springer; 2023

[3] Snapp SS, Blackie MJ, Gilbert RA, Bezner-Kerr R, Kanyama-Phiri GY. Biodiversity can support a greener revolution in Africa. Proceedings of the National Academy of Sciences of the United States of America. 2010;**107**(48):20840-20845

[4] Frasier I, Noellemeyer E, Amiotti N, Quiroga A. Vetch-rye biculture is a sustainable alternative for enhanced nitrogen availability and low leaching losses in a no-till cover crop system. Field Crops Research. 2017;**20**(102):104-112

[5] Wang S, Xiong J, Yang B, Yang X, Du T, Steenhuis TS, et al. Diversified crop rotations reduce groundwater use and enhance system resilience. Agricultural Water Management. 2023;**276**:108067

[6] Stratton AE, Comin JJ, Siddique I, Zak DR, Dambroz Filipini L, Rodrigues Lucas R, et al. Assessing cover crop and intercrop performance along a farm management gradient. Agriculture, Ecosystems and Environment. 2022;**332**:107925

[7] Chahih H, Drevon JJ, Lazali M. Intercropping chickpea with durum wheat enhances nutrient uptake and grain yield under low phosphorus availability. Communications in Soil Science and Plant Analysis. 2024;**55**(12):1733-1746. DOI: 10.1080/00103624.2024.2328618

[8] Xu Z, Li C, Zhang C, Yu Y, van der Werf W, Zhang F. Intercropping maize and soybean increases efficiency of land and fertilizer nitrogen use: A meta-analysis. Field Crops Research. 2020;**246**:107661

[9] Lithourgidis AS, Dordas CA, Damalas CA, Vlachostergios DN. Annual intercrops: An alternative pathway for sustainable agriculture. Australian Journal of Crop Science. 2011;**5**(4):396-410

[10] Seran TH, Brintha I. Review on maize based intercropping. Journal of Agronomy. 2010;**9**(3):135-145

[11] Yilmaz Ş, Atak M, Erayman M. Identification of advantages of maize-legume intercropping over solitary cropping through competition indices in the east mediterranean region. Turkish Journal of Agriculture and Forestry. 2008;**32**(2):111-119

[12] Ma Q, Wu Y, Liu Y, Shen Y, Wang Z. Interspecific interaction and productivity in a dryland wheat/alfalfa strip intercropping. Field Crops Research. 2024;**309**:109335

[13] Chandrasekaran R, Solaimalai A, Jayakumar M, Surendran U. Assessment of intercropping systems and nitrogen levels on nutrient balance and yield of crops under irrigated condition in Western zone of Tamil Nadu. Communications in Soil Science and Plant Analysis. 2023;**54**(19):2675-2682. DOI: 10.1080/00103624.2023.2240369

[14] Hidayat A, Mulyani A. Dryland for agriculture. In: Dryland Management Book towards Productive and

Environmentally Friendly Agriculture. Bogor: Center for Soil and Agroclimatic Research and Development; 2005

[15] Muyassir, Sufardi I, Saputra. Perubahan sifat fisik Inceptisol akibat perbedaan jenis dan dosis pupuk organik. Jurnal Lentera. 2012;**12**(1):1-8

[16] Rusastra IW, Rachman B, Friyatno S. Analysis of the competitiveness and protection structure of Palawija commodities. In: Proceedings of Efficiency and Competitiveness of Farming Systems for Several Agricultural Commodities in Rice Fields. Bogor: Center for Agricultural Socio-Economic Research and Development; 2004

[17] Han F, Javed T, Hussain S, Guo S, Guo R, Yang L, et al. Maize/peanut rotation intercropping improves ecosystem carbon budget and economic benefits in the dry farming regions of China. Journal of Environmental Management. 2024;**353**:120090

[18] Layek J, Das AK, Mitran T, Nath CP, Meena RS, Yadav GS et al. Cereal+Legume Intercropping: An Option for Improving Productivity and Sustaining Soil Health. In: Meena R, Das A, Yadav G, Lal R, editors. Legumes for Soil Health and Sustainable Management. Singapore: Springer; 2018. DOI: 10.1007/978-981-13-0253-4_11

[19] Yadav MR, Singh M, Kumar R, Kumar D, Meena RK, Ram H, et al. Integrated nutrient management in maize-cowpea intercropping system is an attractive option to improve the fodder productivity and quality. Communications in Soil Science and Plant Analysis. 2022;**53**(22):3045-3059. DOI: 10.1080/00103624.2022.2101660

[20] Li L, Zhang F, Li X, Christie P, Sun J, Yang S, et al. Interspecific facilitation of nutrient uptake by intercropped maize and faba bean. Nutrient Cycling in Agroecosystems. 2003;**65**(1):61-71

[21] Ghosh PK, Manna MC, Bandyopadhyay KK, Ajay TAK, Wanjari RH, et al. Interspecific interaction and nutrient use in soybean/sorghum intercropping system. Agronomy Journal. 2006;**98**(4):1097-1108

[22] Beets WC. Multiple Cropping and Tropical Farming Systems. New York: CRC Press; 1982

[23] Center for Soil and Agro-climatic Research and Development. Atlas of Indonesian Agricultural Spatial Planning, Scale 1:1,000,000. Bogor: Center for Soil and Agro-Climatic Research and Development; 2001

[24] Yilmaz Ş, Özel A, Atak M, Erayman M. Effects of seeding rates on competition indices of barley and vetch intercropping systems in the eastern Mediterranean. Turkish Journal of Agriculture and Forestry. 2014;**39**(1):135-143. DOI: 10.3906/tar-1406-155

[25] Palaniappan S. Cropping Systems in the Tropics: Principles and Management. New York: Wiley Eastern Limited; 1985

[26] Banik P, Midya A, Sarkar BK, Ghose SS. Wheat and chickpea intercropping systems in an additive series experiment: Advantages and weed smothering. European Journal of Agronomy. 2006;**24**(4):325-332. DOI: 10.1016/j.eja.2005.10.010

[27] Wu X, Xiao X, Yang Z, Wang J, Steiner J, Bajgain R. Spatial-temporal dynamics of maize and soybean planted area, harvested area, gross primary production, and grain production in the contiguous United States during 2008-2018. Agricultural and Forest Meteorology. 2021;**297**:108240. DOI: 10.1016/j.agrformet.2020.108240

[28] Battisti R, Ferreira MDP, Tavares ÉB, Knapp FM, Bender FD, Casaroli D, et al. Rules for grown soybean-maize cropping system in Midwestern Brazil: Food production and economic profits. Agricultural Systems. 2020;**182**:102850. DOI: 10.1016/j.agsy.2020.102850

[29] Hailu G. A review on the comparative advantage of intercropping systems. Journal of Biology, Agriculture and Healthcare. 2015;**5**(9):28-38. Available from: http://www.iiste.org/Journals/index.php/JBAH/article/view/21387

[30] Ceunfin S, Prajitno D, Suryanto P, Putra ETS. Penilaian kompetisi dan keuntungan hasil tumpangsari jagung kedelai di bawah tegakan kayu putih. Savana Cendana: Jurnal Pertanian Konservasi Lahan Kering. 2017;**2**(1):1-3

[31] Lubis KS, Amalina AN. Land equivalent ratio and competitive index of intercropping vetiver with soybean and organic mulch application. In: IOP Conference Series: Earth and Environmental Science. Vol. 1413. Bristol: IOP Publishing; 2024. p. 012125

[32] Wang Y, Han X, Zhao X, Zhang Y, Qi B, Li L. Grain yield and interspecific competition in an oat-common vetch intercropping system at varying sowing density. Frontiers in Plant Science. 2024;**15**. DOI: 10.3389/fpls.2024.1344110. Available from: https://www.frontiersin.org/journals/plantscience/articles/10.3389/fpls.2024.1344110/full

[33] Deng H, Pan X, Lan X, Wang Q, Xiao R. Rational maize–soybean strip intercropping planting system improves interspecific relationships and increases crop yield and income in the China Hexi oasis irrigation area. Agronomy. 2024;**2024**(14):1220. DOI: 10.3390/agronomy14061220. Available from: https://www.mdpi.com/journal/agronomy

[34] Li Y, Chen P, Fu Z, Luo K, Lin P, Gao C, et al. Maize-soybean relay cropping increases soybean yield synergistically by extending the post-anthesis leaf stay-green period and accelerating grain filling. Crop Journal. 2023;**11**(6):1921-1930. DOI: 10.1016/j.cj.2023.05.011

[35] Khatri N, Dahal KR, Amgain LP, Karki TB. Productivity and economic assessment of maize and soybean intercropping under various tillage and residue levels in Chitwan, Nepal. World Journal of Agricultural Research. 2014;**2**(6A):6-12. DOI: 10.12691/wjar-2-6A-2

[36] Gao Y, Duan A, Qiu X, Liu Z, Sun J, Zhang J, et al. Distribution of roots and root length density in a maize/soybean strip intercropping system. Agricultural Water Management. 2010;**98**(1):199-212. DOI: 10.1016/j.agwat.2010.08.021

[37] Liu J, Yang W. Soybean maize strip intercropping: A solution towards food security in China. Journal of Integrative Agriculture. 2024;**23**(7):2503-2506. DOI: 10.1016/j.jia.2024.02.001

[38] Casali L, Herrera JM, Rubio G. Resilient soybean and maize production under a varying climate in the semi-arid and sub-humid Chaco. European Journal of Agronomy. 2022;**135**:126463. DOI: 10.1016/j.eja.2022.126463

[39] Zustovi R, Landschoot S, Dewitte K, Verlinden G, Dubey R, Maenhout S, et al. Intercropping indices evaluation on grain legume-small grain cereals mixture: A critical meta-analysis review. Agronomy for Sustainable Development. 2024;**44**(1):5

[40] Yang F, Huang S, Gao R, Liu W, Yong T, Wang X. Growth of soybean seedlings in relay strip intercropping

systems in relation to light quantity and red: Far-red ratio. Field Crops Research. 2014;**155**:245-253

[41] Ruswandi D, Supriatna J, Rostini N, Suryadi E. Assessment of sweetcorn hybrids under sweetcorn/chilli pepper intercropping in West Java, Indonesia. Journal of Agronomy. 2016;**15**(3):94-103. DOI: 10.3923/ja.2016.94.103

[42] Kumar N, Singh R, Agrawal RK, Sharma GD, Singh A, Sharma T, et al. Optimizing forage harvest and the nutritive value of Italian ryegrass-based mixed forage cropping under northwestern Himalayan conditions. Frontiers in Plant Science. 2024;**3**(15):1346936

[43] Wahla IH, Ahmad R, Ehsanullah AA, Jabbar A. Competitive functions of component crops in some barley-based intercropping systems. International Journal of Agriculture and Biology. 2009;**11**:69-72. Available from: https://api.semanticscholar.org/CorpusID:86042542

[44] Wang D, Li GR, Zhou BY, Zhan M, Cao CG, Meng QF, et al. Innovation of the double-maize cropping system based on cultivar growing degree days for adapting to changing weather conditions in the North China plain. Journal of Integrative Agriculture. 2020;**19**(12):2997-3012. DOI: 10.1016/S2095-3119(20)63213-0

Section 2

# Climate Adaptation and Stress Tolerance

Chapter 4

# Wild and Cultivated Soybean: Mechanisms of Salt Tolerance and Agricultural Applications

*Deepak B. Shelke, Mahadev R. Chambhare and Ganesh Chandrakant Nikalje*

## Abstract

Soil salinity poses a major challenge to global soybean productivity by inducing osmotic imbalance, ion toxicity, and oxidative stress. Cultivated soybean (*Glycine max*), a glycophytic crop, is particularly salt sensitive, whereas its wild progenitor (*Glycine soja*) exhibits remarkable tolerance through diverse structural and molecular adaptations. This chapter provides a comparative overview of the morphological, physiological, biochemical, and genetic mechanisms underlying salt tolerance in *G. max* and *G. soja*. Under saline conditions, wild soybean maintains superior root architecture, efficient ion exclusion, and sustained photosynthetic performance compared with the cultivated species. Enhanced accumulation of osmolytes and antioxidant enzymes enables *G. soja* to mitigate oxidative damage and preserve cellular homeostasis. At the molecular level, key genes such as *GmSALT3*, *GmNHX5*, *GsCLC-c2*, *GsWRKY40*, *GmFDL19*, and *GsCBRLK* play central roles in ion regulation, ROS detoxification, and ABA-mediated signaling pathways. Transgenic studies confirm that overexpression of these genes improves stress resilience and yield stability under saline conditions. The integration of wild genetic resources with modern molecular breeding and genome-editing tools offers new opportunities to enhance salt tolerance in soybean. Insights from this comparative analysis contribute to developing sustainable strategies for improving crop performance and productivity in salt-affected agroecosystems.

**Keywords:** soybean, salt tolerance, wild relatives, *Glycine max*, *Glycine soja*, transgenic development, molecular breeding

## 1. Introduction

Soybean (*Glycine max* (L.) Merr.) is one of the important oil-yielding legumes globally and belonging to the Fabaceae family. It holds immense significance as economic value not only as a multipurpose crop, serving as a key source of edible oil, protein, and raw material for various food products such as soya milk, tofu, and soya sauce. Owing to its high protein content, soybean forms an essential dietary component for both humans and livestock and also serves as a sustainable feedstock for biodiesel production [1]. Soybean has also been extensively used in biodiesel production. Its demand continues to rise due to its versatility and widespread industrial applications. Currently, soybean contributes nearly two-thirds of globally produced plant protein and ranks among the leading oilseed crops. In response to growing demand, its global production has reached approximately 350 million in recent decades [2]. Rapid climatic fluctuations and environmental degradation have intensified both biotic and abiotic stresses on agriculture. Among the abiotic stresses, drought, cold, and salinity are detrimental to crop productivity. These stresses pose serious threats to global food security by causing progressive decline in yield [1]. To meet future food demands, agricultural production must increase by nearly 70% [3]. Among the various abiotic stresses, salinity is particularly destructive, reducing crop yields and degrading arable lands worldwide. It is projected that by 2050, approximately 50% of the world's arable land could become salinized due to continued soil mismanagement and climate change [2, 4]. Currently, around 1.4 billion hectares—or about 10% of the global land—are affected by salinity, with an additional one billion hectares at risk [5]. Anthropogenic activities—such as excessive fertilizer use, poor irrigation management, and seawater intrusion—further aggravate salinity in both rain-fed and irrigated systems [6, 7].

Salinity imposes osmotic imbalance and ion toxicity in plants, disrupting growth, metabolism, and nutrient homeostasis [8, 9]. Initially, high soil $Na^+$ concentration reduces water uptake, causing osmotic stress; later, ion accumulation in shoots impairs photosynthesis and cellular metabolism [10]. The dominant salts include $Na^+$ and $Cl^-$ ions, which account for 50–80% of the total soluble salts in saline soils [11, 12]. These ions limit the nutrient uptake—especially of $Ca^{2+}$, $K^+$, and $NO_3^-$ and cause oxidative damage to plant tissues.

Soybean, a glycophytic crop, is highly sensitive to elevated soil salinity [13]. Salt accumulation in soil and irrigation water negatively affects plant growth, yield, and physiological attributes. Therefore, developing stress-resilient soybean cultivars has become a global priority. Several soybean genotypes show moderate variation in salt tolerance [1, 2, 13], but progress in breeding salt-resistant varieties has been constrained by the crop's narrow genetic base [14]. Wild relatives of cultivated crops often exhibit greater stress tolerance than domesticated lines. *Glycine soja*, the wild progenitor of *G. max*, demonstrates superior salt tolerance through diverse adaptive mechanisms [2, 15, 16]. Its rich genetic diversity offers valuable traits for breeding salt-tolerant cultivars. Despite several reports describing individual physiological and molecular responses, comprehensive comparative analyses integrating morphological, biochemical, and genetic mechanisms remain limited. This chapter aims to synthesize current understanding of the salt-tolerance mechanisms in cultivated and wild soybean species, identify key adaptive traits and genes contributing to tolerance, and highlight their potential use in breeding and transgenic improvement programs for developing salt-resilient cultivars.

## 2. Habit and Habitat

Soybean (*Glycine max* (L.) Merr.) is an erect, fast-growing, herbaceous leguminous plant that can reach a height of up to 1 m [17]. Native to Asia, soybean is now cultivated globally under diverse agro-climatic conditions [18]. The genus *Glycine* comprises one wild species, one semi-wild species [19], and one cultivated species, along with two polyploid species [20]. Among them, *Glycine max* represents the cultivated soybean, while *Glycine soja* is its wild progenitor. Soybean is one of the most extensively studied oilseed crops, owing to its agronomic and nutritional significance [7]. The wild soybean (*Glycine soja*) shares the same diploid chromosome number (2n = 40) as the cultivated *G. max* and is a rich source of genetic diversity for improving stress resilience and other agronomic traits [17]. Recent studies have reported the existence of polyploid *Glycine* species, such as *G. tabacina* (2n = 80) and *G. tomentella*, which shows variation in chromosome number (2n = 38, 78, or 80) [20, 21]. Wild soybean is native to East Asia, particularly China, Japan, Korea, and Russia, and is typically found in varied habitats such as riverbanks, roadsides, moist areas, and warm climates with fertile, well-drained soils. Globally, the leading producers of soybeans are the United States, Brazil, and Argentina. In India, the major soybean-producing states include Madhya Pradesh, Maharashtra, Rajasthan, Karnataka, and Telangana [22, 23].

## 3. Morphology and anatomy under salt stress

Salt stress significantly influences the morphology and anatomy of soybean plants, affecting growth, vigor, and productivity. The impact varies considerably between cultivated soybean (*Glycine max*) and its wild relative (*Glycine soja*), reflecting differences in their adaptive strategies and genetic makeup.

### 3.1 Cultivated soybean (*Glycine max*)

In cultivated soybean, salinity causes significant reductions in plant height, number of leaves, stem thickness, and both root and shoot biomass [7, 24]. Growth suppression is often associated with inhibited cell expansion, reduced water uptake, and impaired photosynthesis. For example, salt-treated soybean plants generally exhibit shorter hypocotyls, fewer leaves, and lower fresh and dry weights compared to controls, particularly at higher salt concentrations [7, 13].

Anatomically, high salinity leads to notable alterations in the vascular system of soybean roots and stems. Studies have shown a decline in metaxylem vessel number and diameter, restricted cell enlargement, and thickening of the cortex and pith under elevated salt levels [25, 26]. The formation of aerenchyma and thicker epidermal and endodermal layers in roots are adaptive responses that restrict sodium entry and mitigate oxidative damage [27]. Additionally, increased cuticle thickness and reduced stomatal density in leaves act as protective barriers to minimize water loss and ion toxicity. These structural modifications, although beneficial for stress tolerance, often come at the cost of reduced growth and productivity.

### 3.2 Wild soybean (*Glycine soja*)

Wild soybean exhibits more stable growth and structural integrity under salinity compared to its cultivated counterpart. In salt-tolerant lines such as *G. soja* 'Tongyu06311', root length and biomass remain largely unaffected even under high salt concentrations, while sensitive lines like 'Huinan06116' show significant reductions [28]. This resilience is attributed to better root architecture, enhanced vascular organization, and efficient ion compartmentalization.

Anatomical observations indicate that *G. soja* maintains larger cortical and vascular regions with intact xylem structure under salt stress, supporting improved water and nutrient transport. The presence of well-developed aerenchyma and thicker root

| Sr. No. | Trait/response | Cultivated soybean (*Glycine max*) | Wild soybean (*Glycine soja*) | References |
|---|---|---|---|---|
| 1 | Growth | Reduced plant height, leaf number, shoot, and root biomass | Stable growth and biomass under moderate to high salinity | [7, 28] |
| 2 | Root structure | Reduced radicle diameter, and fewer and narrower metaxylem vessels | Well-developed cortex and vascular system, intact metaxylem | [25, 26] |
| 3 | Leaf anatomy | Reduced stomatal density, thicker cuticle, smaller leaf area | Maintains leaf area and stomatal function | [19, 29] |
| 4 | Aerenchyma formation | Present but limited; primarily in high-salt conditions | Prominent and efficient formation across gradients | [27, 28] |
| 5 | Vascular tissue | Collapsed or reduced vessel diameter under high NaCl | Stable and well-differentiated xylem vessels | [26] |
| 6 | Adaptation | Structural restriction and growth inhibition | Sustained growth and ion regulation efficiency | [2, 15] |

**Table 1.**
*Comparative morphological and anatomical responses of cultivated and wild soybean under salt stress.*

cortex enhances its ability to withstand ionic and osmotic imbalances. Comparative studies among *Glycine soja*, *G. tomentella*, and *G. tabacina* further suggest that wild species maintain leaf area and biomass more effectively than cultivated soybean even under increasing salinity levels [29].

A comparison of both species (**Table 1**) highlights that *G. soja* demonstrates special structural adaptations—optimized root architecture, sustained vascular integrity, and efficient ion exclusion—whereas *G. max* primarily responds through growth restriction and tissue thickening. These anatomical and morphological traits collectively contribute to the superior salt tolerance of wild soybean by maintaining ion balance and physiological functionality. Integrating these traits into cultivated genotypes through conventional and molecular breeding could enhance resilience in salt-affected environments.

## 4. Physiological response and salt tolerance

Salt stress exerts profound effects on soybean physiology, influencing photosynthesis, water relations, ion balance, and metabolic activity. The severity of these responses varies across genotypes, salinity levels, and experimental conditions, with cultivated soybean (*Glycine max*) generally exhibiting greater sensitivity compared to its wild relatives (*Glycine soja*), which reveals the adaptive capacities to maintain cellular homeostasis under saline conditions (**Figure 1**; **Table 2**).

### 4.1 Physiological alterations

Salinity reduces the efficiency of photosynthetic processes by disturbing both stomatal and non-stomatal components. Controlled greenhouse and growth chamber

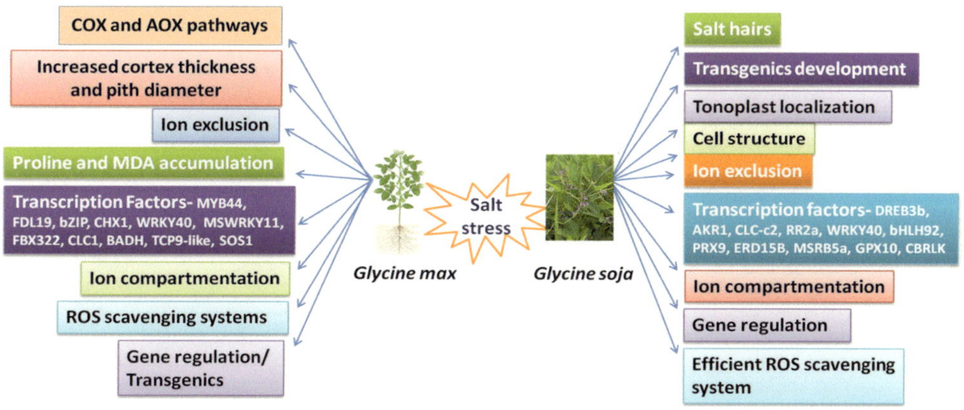


**Figure 1.**
*Strategies involved in the salt tolerance mechanism of cultivated soybean* Glycine max *and wild soybean* Glycine soja. *(AOX—alternative oxidase; COX—cyclooxygenase; ROS—reactive oxygen species; MDA—malondialdehyde; MYB44—myeloblastosis 44; bZIP—basic leucine zipper; CHX1—cation/proton exchanger 1; WRKY40—DNA binding protein 40; FBX322—F-box protein 322, CLC1—chloride voltage-gated channel 1; BADH—betaine aldehyde dehydrogenase; TCP9-like—teosinte branched/cycloidea/proliferating cell factor; SOS1—salt overly sensitive 1; DREB3b—dehydration-responsive element binding; AKR1—Aldo-keto reductase 1; CLC2—chloride voltage-gated channel 2; RR2a, WRKY40—DNA-binding protein 40; bHLH92—basic helix-loop-helix 92; PRX9—peroxidase 9; ERD15—early responsive to dehydration 15; MSRB5a—methionine sulfoxide reductase B; GPX10—glutathione peroxidise 10; CBRLK—calmodulin-binding receptor-like kinase).*

| Sr. No. | Trait/parameter | Cultivated soybean (*Glycine max*) | Wild soybean (*Glycine soja*) | Type of study | References |
|---|---|---|---|---|---|
| 1 | Photosynthetic rate and gas exchange | Strong decline in photosynthetic rate, stomatal conductance, and transpiration under salinity | Maintains higher photosynthetic efficiency and PSI/PSII stability | Controlled pot and growth-chamber studies | [13, 30] |
| 2 | Relative water content (RWC) | Substantial reduction, indicating poor osmotic adjustment | Higher RWC maintained even under moderate salinity | Pot experiments | [31] |
| 3 | Ion regulation ($Na^+$, $K^+$) | High $Na^+$ accumulation, low $K^+/Na^+$ ratio | Efficient $Na^+$ exclusion, maintains higher $K^+/Na^+$ ratio | Controlled and hydroponic assays | [19, 32] |
| 4 | Chlorophyll and pigments | Significant loss of chlorophyll a, b, and carotenoids | Smaller reductions; maintains chlorophyll stability | Greenhouse trials | [33, 34] |
| 5 | Osmolyte accumulation (proline, glycine betaine) | Increases under stress but insufficient to prevent injury | Higher accumulation contributes to osmotic balance | Controlled pot studies | [24] |
| 6 | Antioxidant enzymes (SOD, CAT, POX) | Moderate or inconsistent induction | Strong and sustained up-regulation under salinity | Pot and transgenic studies | [19, 35] |

| Sr. No. | Trait/parameter | Cultivated soybean (*Glycine max*) | Wild soybean (*Glycine soja*) | Type of study | References |
|---|---|---|---|---|---|
| 7 | Oxidative markers (MDA, $H_2O_2$) | High accumulation indicating lipid peroxidation | Low accumulation due to better ROS detoxification | Controlled assays | [30] |
| 8 | Membrane stability/ electrolyte leakage | Elevated leakage and structural damage | Lower leakage; improved membrane integrity | Controlled irrigation trials | [31] |
| 9 | Genetic/ transgenic validation | Overexpression of salt-responsive genes improves tolerance modestly | Wild-derived genes (e.g., *GsWRKY40*, *GsMYB15*) confer strong protection | Transgenic lines (greenhouse and field tests) | [35, 36] |

**Table 2.**
*Comparative physiological and biochemical responses of cultivated and wild soybean under salt stress.*

studies have demonstrated that increased leaf sodium ($Na^+$) accumulation correlates strongly with declines in photosynthetic rate, stomatal conductance, transpiration rate, and water use efficiency [13, 37]. The decrease in carbon assimilation is partly due to stomatal closure caused by osmotic stress and impaired chloroplast function under ion toxicity. Saline irrigation also elevates the electrolyte leakage and reduces relative water content (RWC), indicating membrane damage and reduced water retention capacity. In a pot-based study, the cultivar Giza-111 maintained higher RWC and $K^+/Na^+$ ratio than Giza-35 and Giza-82, suggesting better osmotic adjustment and ion regulation [31]. Similarly, increased NaCl concentrations significantly reduced chlorophyll a, chlorophyll b, and carotenoids in soybean leaves, as observed under controlled experiments using salt-treated irrigation water [33, 34].

Salinity reduces photosynthetic efficiency through both stomatal and non-stomatal limitations. In *G. max*, increased $Na^+$ accumulation leads to stomatal closure, reduced transpiration, and impaired $CO_2$ assimilation. In contrast, *G. soja* sustains higher photosystem stability (PSI/PSII) and maintains a more favorable $K^+/Na^+$ ratio, enabling better osmotic adjustment and chlorophyll retention [30]. The wild species also exhibits lower electrolyte leakage and higher relative water content, indicating improved membrane integrity and water conservation [19]. These physiological responses collectively allow *G. soja* to preserve the photosynthetic performance under saline conditions.

### 4.2 Biochemical adjustments

Soybean plants counteract salt-induced osmotic and oxidative stress through a suite of biochemical responses. Controlled pot experiments and hydroponic assays have shown that salt exposure significantly increases levels of compatible solutes such as proline and glycine betaine, which stabilize cellular proteins and membranes [24]. Concurrently, the accumulation of reactive oxygen species (ROS) triggers enhanced activity of antioxidant enzymes including superoxide dismutase (SOD), peroxidase (POX), and catalase (CAT), helping to scavenge ROS and maintain redox balance.

Increased MDA content under salinity serves as an indicator of lipid peroxidation, reflecting membrane injury. However, wild soybean exhibits lower MDA and $H_2O_2$

accumulation compared to cultivated varieties [30], consistent with its enhanced antioxidant defense. This trend is further supported by biochemical analyses of salt-tolerant lines such as *NARC*-1 and *NARC*-2, which showed higher biological yield and antioxidant activity under 150 mM NaCl stress [37]. These experiments demonstrate that targeted gene manipulation can strengthen intrinsic stress-mitigating biochemical pathways.

### 4.3 Interlinked physiological and biochemical mechanisms

Physiological and biochemical adaptations are closely interconnected in determining overall plant tolerance. In soybean, reduced stomatal conductance limits water loss but also affects $CO_2$ assimilation, which is partly offset by biochemical mechanisms that preserve photosystem efficiency and membrane stability. Osmolyte accumulation (e.g., proline, sugars, glycine betaine) contributes to osmotic balance and protects enzymes, while antioxidant enzymes mitigate ROS-induced damage in chloroplasts and mitochondria. Together, these processes sustain photosynthetic function and cellular homeostasis under salinity stress. The superior salt tolerance observed in wild soybean arises from the synergistic regulation of these physiological and biochemical responses—efficient ion exclusion, enhanced osmotic adjustment, and robust ROS-scavenging mechanisms—making it a valuable genetic resource for improving cultivated soybean resilience through breeding and transgenic approaches [7, 13].

## 5. Gene regulation and salt tolerance

Salt stress in soybean triggers a complex network of molecular responses involving ion transporters, transcription factors, and signaling proteins that collectively maintain ionic balance, osmotic stability, and redox homeostasis. Studies using transcriptomic, quantitative trait loci (QTL), and single-nucleotide polymorphism (SNP) analyses have identified numerous genes contributing to salinity tolerance. These can be broadly classified into four functional categories: (1) ion homeostasis, (2) transcriptional regulation, (3) reactive oxygen species (ROS) detoxification, and (4) abscisic acid (ABA)-mediated signaling pathways. The key salt responsive genes with their primary function and localization are given in **Table 3**.

### 5.1 Ion homeostasis and transport regulation

Maintaining intracellular ionic balance is essential under salinity stress. Soybean achieves this through the coordinated activity of $Na^+/H^+$ exchangers, chloride channels, and potassium transporters. The comparative transcriptomic studies in soybean revealed that genes such as *GmHAK5, GmKUP6*, and *GmCHX20a* are upregulated under saline conditions to regulate $K^+$ transport and maintain a favorable $K^+/Na^+$ ratio [46]. The *GmNHX5* gene, localized to the Golgi apparatus, enhances $K^+$ retention and contributes to $Na^+$ sequestration in vacuoles [38]. Quantitative trait loci (QTL) mapping has identified *GmSALT3* and *GmSALT18* as the major salt tolerance genes derived from *Glycine soja*, both involved in long-distance $Na^+$ and $Cl^-$ exclusion from leaves [39]. Similarly, chloride channel genes (*GmCLC1, GsCLC-c2*, and *CLC-b* family members) facilitate $Cl^-$ compartmentalization, preventing toxic accumulation in shoots [40, 49]. Collectively, these transport-related genes play central roles in sustaining ion homeostasis and minimizing ionic toxicity under saline conditions.

| Sr. No. | Functional category | Gene(s) | Source/localization | Primary function | Experimental evidence | References |
|---|---|---|---|---|---|---|
| 1 | Ion homeostasis | *GmNHX5*, *GmSALT3*, *GmSALT18*, *GmCLC1*, *GsCLC-c2* | Golgi, tonoplast, plasma membrane | $Na^+/H^+$ exchange, $Cl^-$ sequestration, ion balance | QTL mapping, transcriptomics | [38–40] |
| 2 | Transcriptional regulation | *GmFDL19*, *GsWRKY40*, *GsbHLH92*, | Nucleus | Activation of stress genes, ABA signaling, ROS regulation | Transgenic & controlled studies | [35, 36, 41, 42] |
| 3 | ROS detoxification | *GsPRX9*, *GsMSRB5a*, *GsoGPX10.1* | Cytosol, chloroplast, mitochondria | ROS scavenging, lipid peroxidation reduction | Gene overexpression, biochemical assays | [43–45] |
| 4 | ABA and signaling pathways | *GmOST1/SnRK2.6*, *GmERF98*, *GsCBRLK*, *GsRR2a* | Plasma membrane, cytoplasm | ABA signaling, $Ca^{2+}$-mediated stress sensing | Transcriptomic & functional assays | [46–48] |

**Table 3.**
*Summary of functional category, localization, primary function, and experimental evidences of key salt-responsive genes in soybean.*

### 5.2 Transcriptional regulation

Transcription factors (TFs) serve as master regulators that activate stress-responsive genes. Members of the *WRKY, DREB, MYB*, and *bZIP* families have been extensively characterized in soybean. For instance, *GmFDL19*, a basic leucine zipper (*bZIP*) transcription factor, enhances tolerance to drought and salinity through ABA-mediated pathways and promotes early flowering [35]. The *DREB3b* allele from *G. soja* confers greater salt tolerance than the reference *DREB3* gene in cultivated soybean [41]. The WRKY family is particularly important in oxidative stress and hormonal regulation. Co-overexpression of *GsWRKY40* and *GsbHLH92* from *G. soja* significantly enhances salt tolerance while reducing ROS accumulation, demonstrating the synergistic regulation of stress-responsive networks [35]. Similarly, overexpression of *AtWRKY11* in transgenic soybean lines improved survival and growth under saline conditions [42].

### 5.3 ROS detoxification and antioxidant defense

Salt stress leads to oxidative damage through the overproduction of reactive oxygen species (ROS), including superoxide radicals ($O_2^-$), hydrogen peroxide ($H_2O_2$), and hydroxyl ions ($OH^-$). Soybean counteracts this *via* genes encoding antioxidant enzymes and ROS-regulating proteins.

Key examples include *GsPRX9* (peroxidase) and *GsMSRB5a* (methionine sulfoxide reductase), both isolated from *G. soja*, which enhance enzymatic detoxification and regulate ROS signaling [43, 44]. The *GsoGPX10.1* gene, encoding glutathione peroxidase, shows continuous upregulation under salt exposure, highlighting its function in maintaining redox balance [45].

### 5.4 ABA signaling and stress perception pathways

The abscisic acid (ABA) pathway plays a pivotal role in integrating external salt signals with internal physiological responses. Genes such as *GmOST1/SnRK2.6* and *GmERF98* are upregulated under salinity, participating in ABA-dependent stomatal regulation and downstream stress signaling [46]. In wild soybean, *GsCBRLK* (a calcium/calmodulin-binding receptor-like kinase) positively regulates tolerance to both salt and ABA stress by mediating $Ca^{2+}$-dependent signaling cascades [47]. Likewise, *GsRR2a*, a cytokinin response regulator, shows dual responsiveness to salt and alkali stress, suggesting its involvement in multiple hormonal pathways [48].

In summary, the gene regulation in soybean under salinity stress represents a tightly coordinated molecular system. Wild soybean (*G. soja*) contributes to unique alleles and transcription factors that enhance tolerance by fine-tuning ion balance, maintaining ROS homeostasis, and activating ABA-linked signaling networks. The integration of QTL mapping, transcriptomic analysis, and transgenic validation provides valuable insights for molecular breeding and the development of salt-resilient soybean cultivars.

## 6. Development of transgenics for improved salt tolerance

The development of transgenic soybean lines represents a promising strategy to enhance salt tolerance beyond the limits of conventional breeding. Traditional breeding efforts are often constrained by the narrow genetic base of cultivated *G. max*, whereas transgenic approaches allow direct incorporation of stress-responsive genes from wild soybean (*G. soja*), halophytes, and other plant species [39, 41, 43]. Advances in *Agrobacterium*-mediated transformation, genome editing, and transcriptomic analyses have greatly accelerated gene discovery and functional validation.

Several successful transgenic interventions in soybean have focused on maintaining ionic balance under saline conditions. Overexpression of *GmCHX1*, a cation/$H^+$ antiporter, confers enhanced $Na^+$ exclusion and $Cl^-$ sequestration by coordinating root and shoot ion transport [50]. Similarly, *GmNHX5*, localized in the Golgi apparatus, increases intracellular $K^+$ retention and improves the $K^+/Na^+$ ratio in transgenic plants [38]. The introduction of the *BADH* (betaine aldehyde dehydrogenase) gene from *Atriplex hortensis* (*AhBADH*) *via Agrobacterium*-mediated transformation led to stable enhancement of salt tolerance in soybean. Transgenic lines (e.g., TL7) demonstrated better growth, higher chlorophyll content, and reduced ROS accumulation under 300 mM NaCl compared to controls [51]. These greenhouse and field studies confirmed *AhBADH*-expressing lines as the strong candidates for commercial breeding programs.

Transcription factors derived from *Glycine soja* and other species have proven effective in enhancing tolerance by activating multiple downstream stress-response pathways. Overexpression of *GsMYB15* in transgenic *Arabidopsis* enhanced resistance to both salinity and insect stress, highlighting its dual role in defense regulation [52]. Similarly, *GsWRKY40* and *GsbHLH92* co-overexpression in *G. soja* led to greater salt tolerance through suppression of ROS accumulation and stimulation of antioxidant enzyme activities [36]. Transgenic soybean plants expressing *GmFDL19* (a bZIP transcription factor) showed improved drought and salt tolerance through ABA-dependent signaling and early flowering [35]. In addition, *AtMYB44* expression in soybean enhanced drought and salinity tolerance without negative effects on yield [53]. These examples demonstrate that transcription factor engineering provides broad-spectrum stress resilience.

Manipulation of ROS-scavenging systems offers another effective strategy for developing salt-tolerant soybean. Overexpression of *GsPRX9* (peroxidase) and *GsERD15B* (early responsive to dehydration gene) from wild soybean improved the oxidative stress tolerance by maintaining membrane integrity and reducing MDA accumulation [43, 54]. Similarly, *GsMSRB5a*, encoding methionine sulfoxide reductase, enhanced salt and alkali tolerance by regulating ROS signaling [44].

Recent genome-wide studies in *G. soja* identified multiple glutathione peroxidase genes (*GsoGPX10.1, GsoGPX4.2, GsoGPX7.1*), among which *GsoGPX10.1* was most responsive under 250 mM NaCl, highlighting its regulatory role in ROS homeostasis [45]. These transgenic findings, primarily validated under controlled greenhouse conditions, collectively emphasize the potential of ROS-related genes in mitigating oxidative damage during salinity stress.

Several signaling-related genes have been exploited to enhance stress perception and response. The *GsCBRLK* gene, encoding a calcium/calmodulin-binding receptor-like kinase from *G. soja*, improved salt and ABA tolerance when overexpressed in *Medicago sativa* [47]. This gene functions as a positive regulator of $Ca^{2+}$-mediated signaling pathways. Similarly, *GmNFYA13*, encoding a nuclear factor Y subunit, increased salt and drought tolerance in transgenic soybean by enhancing ABA-responsive gene expression [55]. Incorporating multiple genes within ABA and calcium pathways, such as *AtAVP1* (vacuolar $H^+$-pyrophosphatase) and *AtNHX1* ($Na^+/H^+$ antiporter) in co-expression systems, has also produced the additive benefits for salt tolerance by improving ion compartmentalization and osmotic adjustment [56].

| Sr. No. | Gene(s) | Source species | Mechanism of action | Effect in transgenic soybean | Reference |
|---|---|---|---|---|---|
| 1 | *AhBADH*, *GmNHX5*, *GmCHX1* | *Atriplex hortensis*, *G. max* | Osmolyte synthesis, $Na^+$ sequestration, $K^+$ retention | Improved growth, higher chlorophyll, better $K^+/Na^+$ ratio | [38, 50, 51] |
| 2 | *GsWRKY40*, *GsbHLH92*, *GmFDL19*, *GsMYB15*, *AtMYB44* | *G. soja*, *A. thaliana* | Activation of salt-responsive genes, ABA regulation | Enhanced tolerance, reduced ROS, stable yield | [35, 36, 52, 53] |
| 3 | *GsPRX9*, *GsERD15B*, *GsMSRB5a*, *GsoGPX10.1* | *G. soja* | ROS scavenging, membrane protection | Lower MDA and $H_2O_2$, improved survival | [43–45, 54] |
| 4 | *GsCBRLK*, *GmNFYA13*, *AtAVP1*, *AtNHX1* | *G. soja*, *G. max*, *A. thaliana* | ABA-responsive regulation, $Ca^{2+}$-mediated signaling | Enhanced salt and drought resilience | [47, 55, 56] |
| 5 | *SpRAB18* | *Sesuvium portulacastrum* | Stress-induced protein expression, osmoprotection | Increased biomass, photosynthetic stability | [57, 58] |

**Table 4.**
*Key transgenic developments in soybean.*

Beyond soybean relatives, genes from halophytic species have provided promising tools for engineering salinity tolerance. For instance, overexpression of the *SpRAB18* gene from the halophyte *Sesuvium portulacastrum* enhanced growth, chlorophyll content, and antioxidative defense in transgenic soybean under saline conditions [57]. Such cross-species gene transfers expand the genetic toolkit for crop improvement. The key transgenic soybean development studies are enlisted in **Table 4**.

## 7. Conclusion and future prospects

Salt stress remains a major constraint on global soybean productivity, particularly due to the crop's classification as a moderately salt-sensitive glycophyte. Traditional breeding has made limited progress owing to the narrow genetic base of cultivated soybean (*G. max*). In contrast, its wild relative, *G. soja*, exhibits remarkable salt tolerance owing to a wide array of adaptive morphological, physiological, biochemical, and molecular traits. This genetic diversity has emerged as a vital resource for developing stress-resilient cultivars. Significant progress has been made in identifying and functionally characterizing salt-responsive genes from *G. soja*. Several transcription factors (*GsMYB15*, *GsWRKY40*, *GsERD15B*, *GsCBRLK*, *GmFDL19*), antioxidant-related genes (*GsPRX9*, *GsoGPX10.1*, *GsMSRB5a*), and ion transporters (*GmCHX1*, *GsCLC-c2*, *GmCLC1*, *GmNHX5*, *GmSALT3*) have been found to enhance salt tolerance through diverse mechanisms including ROS detoxification, ion homeostasis, and hormone signaling. Many of these genes have been successfully overexpressed in model and crop plants, leading to improved physiological responses and yield stability under saline conditions. Transgenic approaches using genes from *G. soja* and other halophytes (e.g., *AhBADH* from *Atriplex hortensis*) have demonstrated considerable success in greenhouse and early field trials. Notably, lines such as TL7 expressing *AhBADH* have shown stable salt tolerance and promising agronomic traits, highlighting the translational potential of these genes in crop improvement programs. Furthermore, the co-expression of multiple stress-related genes has been more effective than single-gene strategies in enhancing salt tolerance. Despite these advances, large-scale field validation, regulatory clearance, and public acceptance remain critical bottlenecks in the commercial deployment of transgenic salt-tolerant soybean varieties. Continued research into the genetic mechanisms of salt tolerance in *G. soja*, coupled with modern genome-editing and gene-stacking technologies, holds great promise for developing robust soybean cultivars suitable for saline environments. This integrated approach is essential for ensuring future food and energy security under increasing soil salinization and climate variability.

## Acknowledgements

The authors acknowledge the use of ChatGPT (GPT-5, developed by OpenAI) for language polishing of the manuscript.

## Conflicts of interests

All authors have no conflict of interest.

## Author details

Deepak B. Shelke[1], Mahadev R. Chambhare[1] and Ganesh Chandrakant Nikalje[2]*

1 Department of Botany, Amruteshwar Arts, Commerce and Science College, Pune, MS, India

2 Department of Botany, Seva Sadan's R. K. Talreja College of Arts, Science and Commerce, Affiliated to University of Mumbai, Thane, India

*Address all correspondence to: gcnikalje@gmail.com

## References

[1] Shelke DB, Chambhare MR, Nikalje GC, Nikam TD. Improvement of soybean crop for yield, stress tolerance, and value added products using a transgenic approach. Advances in Agriculture. 2023;**2023**:1-26. DOI: 10.1155/2023/8166928

[2] Ly T, Thanh L, Lukasz K, Timothy DC, Kadambot HMS. Superior salt tolerance in wild soybean (*G. soja*) is associated with better ion 'exclusion' ability from leaves and mesophyll cells than cultivated soybean genotypes (*G. max*). Environmental and Experimental Botany. 2023;**211**:105348. DOI: 10.1016/j.envexpbot.2023.105348

[3] van Dijk M, Morley T, Rau ML, et al. A meta-analysis of projected global food demand and population at risk of hunger for the period 2010-2050. Nature Food. 2021;**2**:494-501. DOI: 10.1038/s43016-021-00322-9

[4] Muhammad A, Salah ESH, Salem SA, Hussein HM, Muhammad AK, Muhammad UH, et al. Potential breeding strategies for improving salt tolerance in crop plants. Journal of Plant Growth Regulation. 2023;**42**:3365-3387. DOI: 10.1007/s00344-022-10797-w

[5] Food and Agriculture Organization (FAO). FAO launches first Global Assessment of Salt-Affected Soils in 50 Years. Rome, Italy: FAO; 2024. Available from: https://www.fao.org/newsroom/detail/fao-launches-first-major-global-assessment-of-salt-affected-soils-in-50-years/en

[6] Shelke DB, Chambhare MR, Sonawane HB, et al. Synergistic approaches in halophyte-microbe interactions: Mitigating soil salinity and industrial contaminants for sustainable agriculture. Discover Life. 2025;**55**:11. DOI: 10.1007/s11084-025-09688-3

[7] Shelke DB, Nikalje GC, Chambhare MR, Zaware BN, Suprasanna P, Nikam TD. Na+ and Cl− induce differential physiological, biochemical responses and metabolite modulations in vitro in contrasting salt-tolerant soybean genotypes. 3Biotech. 2019;**9**:91. Available from: https://pubmed.ncbi.nlm.nih.gov/30800602/

[8] Hussain S, Khaliq A, Tanveer M, Matloob A, Hussain HA. Aspirin priming circumvents the salinity-induced effects on wheat emergence and seedling growth by regulating starch metabolism and antioxidant enzyme activities. Acta Physiologiae Plantarum. 2018;**40**:68-75. DOI: 10.1007/s11738-018-2644-5

[9] Li L, Mao X, Wang J, Chang X, Reynolds M, Jing R. Genetic dissection of drought and heat-responsive agronomic traits in wheat. Plant, Cell & Environment. 2019;**42**:2540-2553 Available from: https://pubmed.ncbi.nlm.nih.gov/31077401/

[10] Muchate NS, Nikalje GC, Rajurkar NS, Suprasanna P, Nikam TD. Plant salt stress: Adaptive responses, tolerance mechanism and bioengineering for salt tolerance. The Botanical Review. 2016;**82**(4):371-406. DOI: 10.1007/s12229-016-9173-y

[11] Shelke D, Nikalje G, Nikam T, Maheshwari P, Punita D, et al. Chloride (Cl−) uptake, transport, and regulation in plant salt tolerance. In: Plant M, editor. Abiotic Stress. Hoboken, NJ, USA: Wiley; 2019. pp. 241-268. DOI: 10.1002/9781119463665.ch13

[12] Shelke D, Nikalje G, Bhusare B, et al. Biochemical and physiological responses of contrasting soybean genotypes to individual ($Na^+$, $Cl^-$) and additive (NaCl) salts. Vegetos. 2025;**26**:1-14. DOI: 10.1007/s42535-025-01331-w

[13] Shelke DB, Pandey M, Nikalje GC, Zaware BN, Suprasanna P, Nikam TD. Salt responsive physiological, photosynthetic and biochemical attributes at early seedling stage for screening soybean genotypes. Plant Physiology and Biochemistry. 2017;**118**:519-528. DOI: 10.1016/j.plaphy.2017.07.013

[14] Kofsky J, Zhang H, Song BH. The untapped genetic reservoir: The past, current, and future applications of the wild soybean (*Glycine soja*). Frontiers in Plant Science. 2018;**9**:949. DOI: 10.3389/fpls.2018.00949

[15] Kapazoglou A, Gerakari M, Lazaridi E, Kleftogianni K, Sarri E, Tani E, et al. Crop wild relatives: A valuable source of tolerance to various abiotic stresses. Plants (Basel). 2023;**12**(2):328. DOI: 10.3390/plants12020328

[16] Shelke DB, Nikalje GC, Sahoo PK. Salt stress responses of glycophytic rice and halophytic rice: Physiological, biochemical, and molecular aspects. In: Verma DK, et al., editors. Rice Science Biotechnological and Molecular Advancements. New Jersey, USA: Apple Academic Press; 2017. pp. 53-68

[17] Singh RJ, Chung GH. Cytogenetics of soybean: Progress and perspectives. The Nucleus. 2007;**50**(3):403. Available from: https://hjar.org/index.php/hjar/article/view/172492

[18] Giller KE, Dashiell KE. *Glycine max* (L.) Merr. Record from Protabase. In: van der Vossen HAM, Mkamilo GS, editors. PROTA (Plant Resources of Tropical Africa). Wageningen, Netherlands; 2007. Available from: http://www.prota4u.org

[19] Liu H, Song J, Dong L, Wang D, Zhang S, Liu J. Physiological responses of three soybean species (*Glycine soja, G. gracilis,* and *G. max* cv. Melrose) to salinity stress. Journal of Plant Research. 2017;**130**:723-733. DOI: 10.1007/s10265-017-0929-1

[20] Hsing Y, Hsieh JS, Cl P, et al. Systematic status of the *Glycine tomentella* and *G. tabacina* species complexes (Fabaceae) based on ITS sequences of nuclear ribosomal DNA. Journal of Plant Research. 2001;**114**: 435-442. DOI: 10.1007/PL00014009

[21] Chauhan A, Pant M, Gautam A, Rani A, Pant K, Rai N. Neglected treasures in the wild: *Glycine tomentella* and *Glycine tabacina*: A mini review. Environment Conservation Journal. 2025;**26**(1):293-301. DOI: 10.36953/ECJ.32202025

[22] Song W, Sun S, Wu T, Yang R, Tian S, Xu C, et al. Geographic distributions and the regionalization of soybean seed compositions across China. Food Research International. 2023;**164**:112364. DOI: 10.1016/j.foodres.2022.112364

[23] Kumari S, Dambale AS, Samantara R, Jincy M, Bains G. Introduction, history, geographical distribution, importance, and uses of soybean (*Glycine max* L.). In: Soybean Production Technology: Physiology, Production and Processing. Singapore: Springer Nature; 2025. pp. 1-17. DOI: 10.1007/978-981-97-8677-0_1

[24] Kokebie D, Enyew A, Masresha G, Fentie T, Mulat E. Morphological, physiological, and biochemical responses of three different soybean (*Glycine max* L.) varieties under salinity stress

conditions. Frontiers in Plant Science. 2024;**15**:1440445. DOI: 10.3389/fpls.2024.1440445

[25] Dolatabadian ASA, Modarres MS, Ghanati F. Effect of salinity on growth, xylem structure and anatomical characteristics of soybean. Notulae Scientia Biologicae. 2011;**3**(1):41-45. DOI: 10.15835/nsb315627

[26] Silva BRS, Batista BL, Lobato AS. Anatomical changes in stem and root of soybean plants submitted to salt stress. Plant Biology. 2021;**23**(1):57-65. DOI: 10.1111/plb.13176

[27] El-Rodeny WM, EL-Okkiah AF. Physiological and anatomical changes in *Glycine max* L. under salinity stress. In: Egypt. International Journal of Botany conferences. Egyptian: Botanical Society; 2012. pp. 37-50

[28] Jiao Y, Bai Z, Xu J, Zhao M, Khan Y, Hu Y, et al. Metabolomics and its physiological regulation process reveal the salt-tolerant mechanism in *Glycine soja* seedling roots. Plant Physiology and Biochemistry. 2018;**126**:187-196. Available from: https://pubmed.ncbi.nlm.nih.gov/29525442/

[29] Kao W, Tyng-Tyng T, Hung-Chieh T, Chen-Ning S. Response of three *glycine* species to salt stress. Environmental and Experimental Botany. 2006;**2006**(56):120-125. DOI: 10.1016/J.ENVEXPBOT.2005.01.009

[30] Yan K, He W, Bian L, et al. Salt adaptability in a halophytic soybean (*Glycine soja*) involves photosystems coordination. BMC Plant Biology. 2020;**20**:155. DOI: 10.1186/s12870-020-02371-x

[31] El Sabagh A, Omar A, Saneoka H, Barutçular C. Comparative physiological study of soybean (*Glycine max* L.) cultivars under salt stress. Journal of Agricultural Sciences. 2015;**25**(3):269-284

[32] Le LTT, Kotula L, Colmer TD, Siddique KH. Superior salt tolerance in wild soybean (*G. soja*) is associated with better ion 'exclusion' ability from leaves and mesophyll cells than cultivated soybean genotypes (*G. max*). Environmental and Experimental Botany. 2023;**211**:105348

[33] Sadak MS, Abd El-Hameid AR, Zaki FS, Dawood MG, El-Awadi ME. Physiological and biochemical responses of soybean (*Glycine max* L.) to cysteine application under sea salt stress. Bulletin of the National Research Centre. 2020;**44**:1-10. DOI: 10.1186/s42269-019-0259-7

[34] Linh N, Cham L, Thang V. Effects of salinity stress on the growth, physiology, and yield of soybean (*Glycine max* (L.) Merrill). Vietnam Journal of Agricultural Sciences. 2021;**4**(2):1043-1055. DOI: 10.31817/vjas.2021.4.2.05

[35] Li M, Man X, Huiying M, Peng F, Tong C, Xiaohuan S, et al. Wild soybean (*Glycine soja*) transcription factor GsWRKY40 plays positive roles in plant salt tolerance. The Crop Journal. 2024;**12**(3):766-775. DOI: 10.1016/j.cj.2024.03.011

[36] Li Y, Chen Q, Nan H, Li X, Lu S, Zhao X, et al. Overexpression of GmFDL19 enhances tolerance to drought and salt stresses in soybean. PLoS One. 2017;**12**(6):e0179554. DOI: 10.1371/journal.pone.0179554

[37] Khalid N, Saeed A. Physiological responses of seven varieties of soybean [*Glycine max* (L.) Merr.] to salt stress. Plant Science Today. 2023;**10**:199-209

[38] Sun T, Ma N, Wang C, Fan H, Wang M, Zhang J, et al. A golgi-localized

sodium/hydrogen exchanger positively regulates salt tolerance by maintaining higher K+/Na+ ratio in soybean. Frontiers in Plant Science. 2021;**12**:638340. DOI: 10.3389/fpls.2021.638340

[39] Guo X, Jiang J, Liu Y, Yu L, Chang R, Guan R, et al. Identification of a novel salt tolerance-related locus in wild soybean (*Glycine soja* sieb. Zucc.). Frontiers in Plant Science. 2021;**12**:791175. DOI: 10.3389/fpls.2021.791175

[40] Wei PP, Che BN, Shen LK, Cui YQ, Wu SY, Cheng C, et al. Identification and functional characterization of the chloride channel gene, GsCLC-c2 from wild soybean. BMC Plant Biology. 2019;**19**(1):121. DOI: 10.1186/s12870-019-1732-z

[41] Hou Z, Li Y, Cheng Y, Li W, Li T, Du H, et al. Genome-wide analysis of DREB genes identifies a novel salt tolerance gene in wild soybean (*Glycine soja*). Frontiers in Plant Science. 2022;**13**:821647

[42] Wang Y, Jiang L, Chen J, Tao L, An Y, Cai H, et al. Overexpression of the alfalfa WRKY11 gene enhances salt tolerance in soybean. PLoS One. 2018;**13**(2):e0192382. DOI: 10.1371/journal.pone.0192382

[43] Jin T, Sun Y, Zhao R, Shan Z, Gai JY, Li Y. Overexpression of peroxidase gene GsPRX9 confers salt tolerance in soybean. International Journal of Molecular Sciences. 2019;**20**:(15):3745. DOI: 10.3390/ijms20153745

[44] Sun X, Sun M, Jia B, Qin Z, Yang K, Chen C, et al. A *Glycine soja* methionine sulfoxide reductase B5a interacts with the Ca2+/CAM-binding kinase GsCBRLK and activates ROS signaling under carbonate alkaline stress. The Plant Journal. 2016;**86**(6):514-529. DOI: 10.1111/tpj.13187

[45] Aleem M, Aleem S, Sharif I, Wu Z, Aleem M, Tahir A, et al. Characterization of SOD and GPX gene families in the soybeans in response to drought and salinity stresses. Antioxidants (Basel). 2022;**11**(3):460. DOI: 10.3390/antiox11030460

[46] Pruthi R, Chaudhary C, Sharma J, Rana P, Kondi RKR, Richards J, et al. A comparative transcriptomic analysis provides insights into molecular mechanisms driving salt tolerance in soybean. Scientific Reports. 2025;**15**(1):31869. Available from: https://pubmed.ncbi.nlm.nih.gov/40883553/

[47] Bai X, Liu J, Tang L, Cai H, Chen M, Ji W, et al. Overexpression of GsCBRLK from *Glycine soja* enhances tolerance to salt stress in transgenic alfalfa (*Medicago sativa*). Functional Plant Biology. 2013;**40**(10):1048-1056

[48] Chen C, Liu A, Ren H, Yu Y, Duanmu H, Duan X, et al. Genome-wide analysis of *Glycine soja* response regulator GsRR genes under alkali and salt stresses. Frontiers in Plant Science. 2018;**9**:1306

[49] Liu X, Liu F, Zhang L, Cheng C, Wei P, Yu B. GsCLC-c2 from wild soybean confers chloride/salt tolerance to transgenic Arabidopsis and soybean composite plants by regulating anion homeostasis. Physiologia Plantarum. 2021;**172**(4):1867-1879. DOI: 10.1111/ppl.13396

[50] Qu Y, Guan RX, Bose J, Henderson SW, Wege S, Qiu LJ, et al. Soybean CHX-type ion transport protein *GmSALT3* confers leaf Na+ exclusion via a root derived mechanism, and Cl-exclusion via a shoot derived process. Plant, Cell & Environment. 2021;**44**(3):856-869. DOI: 10.1111/pce.13947

[51] Yu Z, Niu L, Cai Q, Wei J, Shang L, Yang X, et al. Improved salt-tolerance

of transgenic soybean by stable over-expression of AhBADH gene from *Atriplex hortensis*. Plant Cell Reports. 2023;**42**(8):1291-1310

[52] Shen XJ, Wang YY, Zhang YX, Guo W, Jiao YQ, Zhou XA. Overexpression of the wild soybean R2R3-MYB transcription factor *GsMYB15* enhances resistance to salt stress and *Helicoverpa armigera* in transgenic Arabidopsis. International Journal of Molecular Sciences. 2018;**19**(2018):3958. DOI: 10.3390/ijms19123958

[53] Seo JS, Sohn HB, Noh K, et al. Expression of the Arabidopsis *AtMYB44* gene confers drought/salt-stress tolerance in transgenic soybean. Molecular Breeding. 2012;**29**:601-608. DOI: 10.1007/s11032-011-9576-8

[54] Jin T, Sun Y, Shan Z, He J, Wang N, Gai JY, et al. Natural variation in the promoter of *GsERD15B* affects salt tolerance in soybean. Plant Biotechnology Journal. 2020;**19**:1155-1169

[55] Ma XJ, Fu JD, Tang YM, Yu TF, Yin ZG, Chen J, et al. *GmNFYA13* improves salt and drought tolerance in transgenic soybean plants. Frontiers in Plant Science. 2020;**11**:587244

[56] Nguyen NT, Vu HT, Nguyen TT, et al. Co-expression of Arabidopsis *AtAVP1* and *AtNHX1* to improve salt tolerance in soybean. Crop Science. 2019;**59**:1133-1143

[57] Kulkarni J, Sahoo SA, Herzyk P, Barvkar VT, Kumar SA, Ravichandran J, et al. Early-responsive molecular signatures associated with halophytic adaptation in *Sesuvium portulacastrum* (L.). Plant, Cell and Environment. 2024;**47**(3):961-975. DOI: 10.1111/pce.14767

[58] Tidke SD, Kedar PD, Chambhare MR, Khandagale AS, Dawani RJ. The influence of magnetic fields and magnetized water on the growth and physio-chemical parameters of soybean. Discover Plants. 2025;**2**(1):1-19. DOI: 10.1007/s44372-025-00227-y

Section 3

# Digital and Precision Agriculture Approaches

Chapter 5

# Advancing Soybean Yield Prediction through Remote Sensing: A Review of Spectral Imaging Applications

*Alfadhl Y. Alkhaled*


## Abstract

Soybeans are an essential crop for animal feed, biofuel production, and global food security. Maximizing its yield under different management and environmental conditions is difficult, and this calls for innovative solutions such as remote sensing. This chapter reviews the application of spectral imaging remote sensing in soybean yield prediction, focusing on the integration of different spectral bands (visible: VIS, near-infrared: NIR, and shortwave infrared: SWIR) and vegetation indices (VI) to assess crop performance. Recent advancements in hyperspectral and multispectral imaging, combined with machine and deep learning models, have enabled more precise estimations of soybean yield by incorporating environmental variables such as soil moisture, nitrogen status, and weather conditions. By reviewing state-of-the-art methodologies and case studies, this chapter highlights the benefits and limitations of remote sensing in precision agriculture. It provides insights into how these technologies can support decision-making for sustainable soybean production, improving productivity while minimizing environmental impact. Finally, future research directions and potential improvements in remote sensing-based yield prediction models are discussed, focusing on integrating artificial intelligence (AI) and advanced data analytics.


**Keywords:** remote sensing, soybean yield prediction, spectral imaging, precision agriculture, machine and deep learning, crop monitoring

## 1. Introduction

Soybeans are one of the most important global crops, serving as a key source of protein for animal feed and human consumption, providing vegetable oil, and contributing to sustainable agriculture and biofuel production. According to the Food and Agriculture Organization [1], soybeans account for approximately 65% of the global protein meal market and 27% of the global vegetable oil market. For example, in the United States, soybeans account for over 90% of total oilseed production [2]. Additionally, the U.S. Energy Information Administration [3] reported that soybean oil is a primary feedstock for biodiesel production, making up about 44% of the total biodiesel feedstock. Given the crop's economic and ecological significance, ensuring stable and high yields is vital. In light of the ever-increasing global population and the challenges posed by climate change, accurate and timely soybean yield prediction has become increasingly important for improving agricultural planning, ensuring food security, and enhancing economic profitability [4–6]. However, achieving accurate predictions remains challenging, as various factors highly influence soybean yields, including environmental, agronomic, genomic, and management-related variables [4, 7, 8].

### 1.1 Soybean yield prediction

Soybean yield prediction refers to estimating the quantity of soybean production expected from a specific area based on various influencing factors such as environmental conditions, crop management practices, and input levels [9]. Accurate yield prediction is critical in optimizing agricultural decision-making, enabling farmers to adjust inputs such as fertilization, irrigation, and pesticide application more efficiently and sustainably [8, 10, 11]. For instance, determining the optimal nitrogen requirements for achieving high yields can minimize excessive fertilizer use, thereby reducing the risk of nitrate leaching into groundwater and preventing water contamination [12]. Similarly, applying the appropriate amount of irrigation conserves water and helps reduce plant stress and humidity-related diseases, contributing to overall crop health and environmental protection [13].

Traditional yield estimation methods, such as field sampling and crop growth models, are valuable but face the difficulties of low spatial coverage and slow operation speed and are based on local parameters [9]. These limitations have motivated the agricultural research community to explore more scalable and data-driven approaches. In recent years, remote sensing techniques have become increasingly suggested for nondestructive, real-time monitoring of large agricultural areas with high temporal and spatial resolution [14–20].

Remote sensing has been widely implemented in precision agriculture for several applications, including nutrient status monitoring [21], weed detection [22], disease monitoring [23], and yield prediction [12, 15, 24]. Its effectiveness lies in detecting a wide range of biophysical and biochemical indicators of the crops, such as water status, chlorophyll content, and canopy structure, that are strongly correlated with plant health and productivity. These indicators are derived from spectral reflectance data, which captures how plants reflect light across different regions of the electromagnetic spectrum. By collecting repeated imagery of crops during key phenological stages, researchers can observe changes in reflectance values, particularly in the visible (VIS: 400–700 nm) [25], red-edge (680–750 nm) [23], near-infrared (NIR: 750–1300 nm) [26, 27], and shortwave infrared (SWIR: 1300–2500 nm) [27] regions. These reflectance patterns are directly influenced by plant physiology. For example, healthy leaves absorb strongly in the red and blue at the VIS region and reflect in the NIR due to internal cellular structures. As the crop progresses through its growth stages, shifts in these spectral signatures, driven by changes in pigment concentration, canopy density, and water content, can be quantitatively linked to final yield. This makes spectral imaging indispensable for nondestructive, time-sensitive yield estimation in soybean fields.

To support these applications, remote sensing platforms, ranging from spaceborne (satellites), airborne, unmanned aerial vehicle (UAV), and ground-based, are equipped with a variety of sensors, including red-green-blue (RGB), multispectral, hyperspectral, thermal, and light detection and ranging (LiDAR) [25, 26, 28–32]. Each of these sensors varies significantly in its spatial, spectral, and temporal resolutions, offering distinct agricultural monitoring advantages. RGB sensors capture images in the VIS spectrum and are primarily used to visually evaluate canopy cover, phenotyping, and disease detection [25]. Multispectral sensors measure reflectance across a limited number of bands, including red-edge and NIR, and are widely used to compute vegetation indices (VI) to evaluate plant health and stress [33]. Hyperspectral sensors offer hundreds of narrow bands, enabling detailed analysis of plant biochemical and physiological traits such as chlorophyll, nitrogen, and water content [27]. Thermal sensors detect infrared radiation emitted from the crop surface to estimate canopy temperature, which is closely related to plant water stress and transpiration [34]. LiDAR sensors, using laser pulses, generate high-resolution three-dimensional (3D) models of crop structure and terrain, facilitating accurate measurement of canopy height, density, and biomass [35]. The result is a massive volume of complex, high-dimensional data, which presents both an opportunity and a challenge. Advanced analytical techniques are needed to extract valuable insights from such data, which is where artificial intelligence (AI) comes into play.

The AI encompasses computational techniques that facilitate machines to learn from data and make informed decisions without being explicitly programmed [26, 36]. In yield prediction, AI is mainly implemented through machine and deep learning algorithms. Machine learning methods, such as Random Forest (RF), Support Vector Machines (SVM), and k-Nearest Neighbors (kNN), are effective for structured inputs like VI and environmental parameters [9, 24, 33, 37–39]. Meanwhile, deep learning architectures, including Convolutional Neural Networks (CNN) and Long Short-Term Memory (LSTM) networks, are capable of automatically extracting features from raw spectral imagery and capturing complex spatial-temporal patterns [16, 37, 40–43]. These AI-driven models improve prediction accuracy by modeling the nonlinear and multidimensional relationships between plant characteristics and yield outcomes.

For soybean crops, variability in canopy architecture, phenological development, and groups present unique challenges for accurate yield estimation [10, 44, 45]. Plants from different maturity groups follow distinct allometric trajectories, and their leaf senescence patterns are often asynchronous. These physiological differences significantly affect spectral reflectance signatures, complicating the relationship between VI and actual yield. As a result, predictive models must account for this heterogeneity to avoid overgeneralization and improve robustness across diverse growing conditions. However, these challenges also highlight the critical role of timely and precise remote sensing data, especially from spectral imaging platforms capable of capturing subtle physiological changes across growth stages [11]. When combined with environmental and management variables, such as soil properties, moisture levels, and fertilization regimes, remote sensing enables a holistic approach to yield prediction, supporting more informed, site-specific decision-making in precision agriculture.

This review focuses on the current state of remote sensing applications for soybean yield prediction, emphasizing spectral imaging techniques, key sensor platforms, and modeling approaches. By synthesizing recent research findings, this chapter aims to inform future efforts to enhance prediction accuracy, integrate AI-based analytical tools, and promote more sustainable, scalable, and data-driven soybean production systems under diverse environmental and management conditions.

### 1.2 Search methodology

This study used specific keywords to search the Scopus database, which was selected for its comprehensive coverage of peer-reviewed literature in agricultural and environmental sciences. The keyword combination used was as follows: ("remote sensing" OR "hyperspectral imaging system" OR "multispectral imaging" OR "RGB" OR "thermal imaging" OR "LiDAR imaging") AND ("yield prediction") AND ("soybean"), focusing primarily on articles published between 2020 and 2025, with select earlier works included for their contextual relevance. The initial search retrieved a total of 102 studies. After careful review to ensure that the selected articles specifically addressed soybean and remote sensing as the central focus, 40 studies were excluded due to differing research scopes or unrelated methodologies. The final selection included 62 relevant studies, forming the basis of this review.

## 2. Fundamentals of remote sensing for crop monitoring

### 2.1 Overview of remote sensing platforms

Remote sensing platforms are systems that serve as carriers to support and transport various types of sensors that acquire spectral imagery across different spatial, temporal, and spectral resolutions and varying observational scales. Depending on the intended application, these platforms may support hyperspectral, multispectral, thermal, or other sensor types. As illustrated in **Figure 1**, remote sensing platforms are generally categorized into four types: spaceborne, aircraft, UAV (unmanned aerial vehicles), and ground-based systems [10]. The main difference among these platforms is the altitude at which they operate, which directly affects both the area observed and the pixel size, the ground area represented by each pixel in the image. As platform altitude increases, the observation area becomes larger, but the pixel size increases, resulting in lower spatial resolution [18, 35].

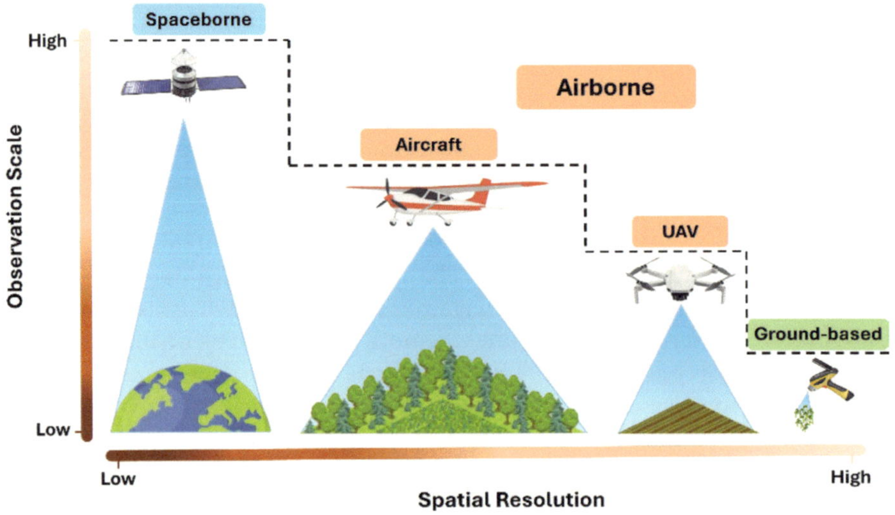


**Figure 1.**
*Remote sensing platforms by altitude and resolution: spaceborne, airborne, UAV-based, and ground-based systems. Lower altitudes provide higher spatial resolution but reduced coverage area.*

Spaceborne remote sensing platforms are satellite-based systems that orbit the Earth and capture data across broad geographic areas [36, 42]. These platforms are equipped with sensors that operate either actively, by transmitting their signals and analyzing the reflected responses, or passively, by relying on sunlight or naturally emitted energy from the Earth's surface. Optical remote sensing, a type of passive sensing, gathers reflected solar radiation and is commonly used in agricultural and environmental monitoring. Satellite imagery is best suited for monitoring large-scale applications that require consistent data collection over time and large spatial areas (e.g., > 100 $km^2$) (**Table 1**). However, both spatial resolution (typically >10 meters per pixel) and temporal resolution (revisiting times ranging from several days to weeks) are often limited, and data acquisition can be affected by atmospheric conditions like cloud cover. With the increasing availability of both public and commercial satellite missions, researchers can now have access to a wide range of datasets, from moderate-resolution imagery (~7–8 m) provided by missions such as Sentinel-2 and Landsat to high-resolution imagery (<5 m) available through platforms like PlanetScope, depending on specific application needs and cost constraints [7, 44, 46].

Airborne remote sensing platforms, such as aircraft and UAVs, carry advanced sensors capable of capturing high-resolution aerial imagery, including multispectral, hyperspectral, and LiDAR technologies [25, 27, 29, 33]. While aircraft can cover extensive areas (ranging from tens to hundreds of square kilometers) and typically provide spatial resolutions finer than 5 cm, UAVs generally operate at lower altitudes, delivering imagery with spatial resolutions between 1 cm and 20 cm over smaller areas (typically less than 20 $km^2$). UAV images are often calibrated using reference targets captured during the same flight to ensure accuracy. However, changes in environmental conditions, such as lighting and cloud cover, can introduce inconsistencies in spectral measurements if not adequately addressed [30].

Lastly, ground-based remote sensing platforms are primarily used to collect high-resolution data for detailed analysis of land surface features [9]. These platforms are instrumental in exploring the interactions between signal-object and signal-sensor [45]. Based on how sensors are mounted, ground-based systems can take various forms, including handheld units, tripods, stationary towers, mobile platforms mounted on robots or tractors, and crane-supported devices. Typically, these platforms are utilized for close-range sensing applications, such as monitoring individual plants' physiological or spectral characteristics or small experimental plots, like a single soybean plant or a limited section of a soybean field. A comparative summary of the different remote sensing platforms, highlighting their spatial, spectral, and temporal resolution, operational altitude, area coverage, flight duration, advantages, and limitations, can be found in **Table 1**. Overall, remote sensing for soybean yield prediction in commercial and research fields in precision agriculture often relies on airborne or ground-based platforms due to their high resolution and lower cost. Spaceborne platforms are currently less practical for this purpose, as higher resolution is needed to establish strong correlations between spectral data and yield. This can be improved shortly due to the improvement of space technologies and sensors mounted on satellite platforms, which can produce high-resolution images.

## 2.2 Overview of remote sensing sensors

Remote sensing sensors are crucial in advancing precision agriculture and smart farming. These technologies effectively monitor crop status and environmental conditions without physical contact, enabling large-scale data collection with high spatial

| Parameter | Spaceborne | Aircraft | UAV | Ground-based |
|---|---|---|---|---|
| Spatial resolution | Low (>10 m) | Medium (0.2–2 m) | High (1–20 cm) | Very High (< 1 cm) |
| Spectral resolution | Low | Medium to high | Very High | Very High |
| Temporal resolution | Days | Hours | Minutes | Minutes to hours |
| Altitude above ground | 700–36,000 km | 0.3–12 km | Up to 150 m | 1–2 m |
| Area coverage | Very Large | Large | Medium | Small |
| Sensor types | Multispectral | Hyperspectral | Multispectral/ Hyperspectral | Multispectral/ Hyperspectral |
| Key advantages | • Large-area coverage<br>• Long-term monitoring<br>• Free data available (e.g., Sentinel, Landsat) | • Flexible sensor payloads<br>• Moderate spatial coverage | • Low operational cost<br>• High flexibility<br>• Frequent deployment | • High spatial accuracy<br>• Low setup cost<br>• Real-time measurements |
| Key limitations | • Affected by cloud cover<br>• Fixed revisit times | • Expensive operation<br>• Weather-dependent<br>• Complex logistics | • Limited flight time<br>• Affected by wind/cloud<br>• Small coverage | • Labor-intensive<br>• Limited spatial extent |

**Table 1.**
*Comparative characteristics of remote sensing platforms relevant to agricultural monitoring, highlighting their technical capabilities and operational constraints.*

and temporal resolution. Various sensor types, each with distinct spectral, spatial, and operational characteristics, are applied in agricultural applications to support tasks such as disease detection, nutrient monitoring, weed identification, structural analysis of crops, and yield prediction [16, 21–23]. Understanding the differences between these sensors is essential for selecting the most appropriate system for specific agricultural goals. **Figure 2** shows five commonly used remote sensing sensor types in agriculture: RGB, multispectral, hyperspectral, thermal, and LiDAR. These sensors differ in spectral characteristics, data acquisition methods, and suitability for monitoring objectives.

The most used sensors in modern agriculture are optical sensors, which include RGB, multispectral, and hyperspectral imaging systems [6, 25, 32]. RGB sensors capture images in the VIS spectrum (red, green, and blue), offering true-color visuals that resemble what the human eye perceives. They are widely used for general plant monitoring, visual classification, and seasonal change tracking. Multispectral sensors extend beyond the VIS spectrum, including bands like red-edge and NIR. These sensors commonly calculate VI, such as the Normalized Difference Vegetation Index (NDVI) and Enhanced Vegetation Index (EVI), which help assess plant health, canopy vigor, and photosynthetic efficiency. Hyperspectral sensors, on the other hand, provide high-resolution data across hundreds of contiguous spectral bands. This level of spectral detail allows for the precise detection of nutrient deficiencies, early-stage disease symptoms, and subtle chemical changes within plant tissues. However, hyperspectral imaging requires more advanced analytical techniques and larger data storage capacities.

**Figure 2.**
*Overview of standard remote sensing sensors based on spectral characteristics: RGB, multispectral, hyperspectral, thermal, and LiDAR. Each category is illustrated with its typical data output and example sensor type, highlighting the diversity of agricultural and environmental monitoring technologies.*

Unlike passive optical sensors that depend on sunlight reflectance, active sensors such as light detection and ranging (LiDAR) and thermal infrared systems generate their energy source [35]. LiDAR uses laser pulses to measure the time the light returns after hitting an object. This enables the creation of detailed 3D models of terrain and vegetation structures, such as tree height, canopy density, and surface topography. LiDAR is particularly valuable in biomass estimation and structural analysis of crop fields. Thermal sensors detect the infrared radiation emitted by surfaces, such as soil or plant leaves [34]. These sensors are often used to evaluate plant water stress by identifying changes in leaf temperature, which may signal reduced transpiration or water deficiency. While thermal imaging provides critical insights for irrigation planning and stress detection, it is sensitive to environmental conditions and requires precise calibration for accurate measurements.

## 2.3 Understanding crop reflectance

As explained in the previous section, remote sensors generate 3D spectral image cubes, consisting of two spatial dimensions (x and y) and one spectral dimension ($\lambda$), which represent reflectance values across a range of wavelengths. These 3D image cubes are often converted into two-dimensional (2D) spectral profiles for easier visualization and analysis, as shown in **Figure 3**. The solid black line in this figure represents a typical leaf reflectance spectrum obtained using a hyperspectral sensor,

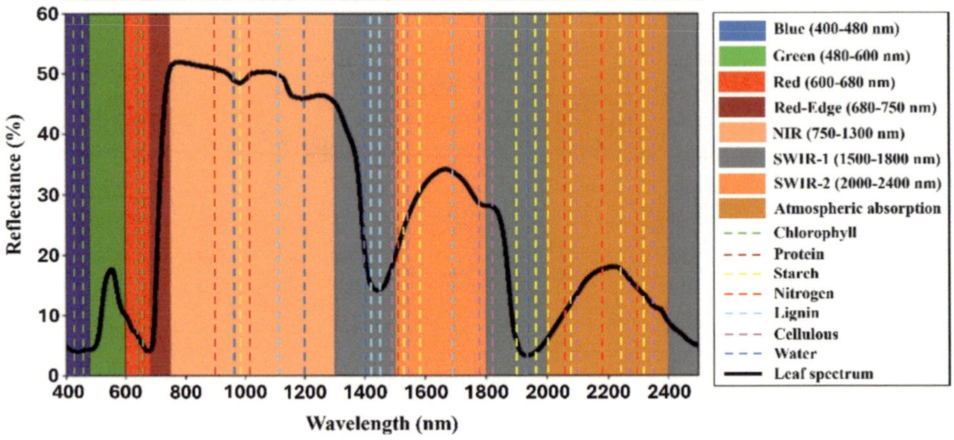


**Figure 3.**
*The leaf reflectance spectrum shows absorption features associated with foliar chemical constituents. The original spectral data was adapted from Ref. [47]; the illustration was modified by the author.*

which captures data from about 400 to 2500 nm across approximately 480 narrow bands. The x-axis shows the wavelength in nanometers (nm), while the y-axis displays reflectance percentage. Reflectance refers to the amount of sunlight reflected off a surface, unlike transmittance (light passing through) or absorption (light absorbed by the material). Reflectance is commonly used in remote sensing for precision agriculture because it provides essential information about crop health and condition (chemistry, morphology, and structure).

In general, spectral profiles obtained from vegetation can include several wavelength regions, depending on the number and range of bands captured. These regions include the VIS (400–680 nm), red-edge (680–750 nm), NIR (750–1300 nm), and SWIR (1300–2500 nm) (**Figure 3**). Each region is linked to the plant's specific chemical and structural properties. The VIS region reflectance in the blue (400–480 nm), green (480–600 nm), and red (600–680 nm) wavelengths is influenced mainly by leaf pigments like chlorophyll (430–460 nm), carotenoids (450 nm), and anthocyanins (550 nm). Healthy plants absorb more light in the blue and red bands for photosynthesis, which explains why leaves appear green (due to higher reflection in that band) [23]. The narrow red-edge region marks a sharp increase in reflectance, transitioning from the red to the NIR region. This region is susceptible to chlorophyll content and can indicate changes in plant health [26]. A shift in the red-edge position toward SWIR may suggest plant stress or reduced chlorophyll levels. The reflectance in the NIR region is determined mainly by internal leaf structure and canopy arrangement. High reflectance is typically a sign of healthy leaf tissue. Also, water and biochemical constituents such as proteins, starch, and lignin can affect NIR reflectance; however, water content can sometimes mask those signals [10]. Lastly, the SWIR region is susceptible to water content. As leaf water increases, reflectance in the SWIR region decreases. It also provides information on traits like nitrogen, starch, and lignin [11]. However, bands in areas like 1300–1500 nm, 1800–2000 nm, and 2400–2500 nm are often removed from analysis because of atmospheric water absorption and weak solar signals. Understanding how plants reflect solar radiation across these regions is essential for using remote sensing effectively in agriculture (see **Figure 3**).

### 2.4 Spectral features and vegetation indices

Spectral data acquired from remote sensing systems typically span a broad range of wavelengths, from approximately 400 nm (VIS light) to 2500 nm (SWIR), as shown in **Figure 3**, depending on the sensor type and the study's specific objectives. Due to the high dimensionality of spectral data, which can include hundreds of spectral bands, researchers have developed several strategies to process, reduce, and interpret this information efficiently. One common approach is to divide the spectrum into key regions, such as the VIS, NIR, and SWIR bands. These regions are associated with different plant physiological responses; for example, the VIS region reflects chlorophyll content, the NIR region is sensitive to canopy structure, and the SWIR region relates to water content in vegetation. Another widely used method is feature selection, in which researchers identify the most informative spectral bands or combinations of bands that correlate with ground-truth measurements, such as leaf nitrogen content, disease severity, or yield data. Instead of analyzing all spectral bands (e.g., 480 bands in a hyperspectral image), only a subset, sometimes as little as 10% of the total, may be used to build efficient and accurate prediction models [21]. These selected features can be distributed across the entire spectral range and are often chosen based on statistical correlation, regression analysis, or machine learning techniques. In addition to raw spectral bands and feature selection, researchers frequently rely on VI to interpret spectral data. The VIs are mathematical combinations of two or more spectral bands designed to highlight specific biophysical properties of plants. For instance, the NDVI uses red and NIR bands to estimate plant vigor and chlorophyll content, while indices such as NDWI assess water content and PRI detect photosynthetic activity. These indices are especially useful for monitoring plant health, detecting stress, estimating biomass, and guiding irrigation decisions.

## 3. Remote sensing applied in soybean yield prediction

Recent studies on soybean yield prediction using remote sensing have primarily employed three different platforms: satellite, UAV, and ground-based. As shown in **Table 2**, approximately half of the studies, around 52%, have utilized satellite-based remote sensing, highlighting it as the most used platform. The widespread adoption of satellite imagery is mainly due to its ability to cover vast agricultural areas efficiently and to provide consistent, repeatable temporal resolution throughout the growing season. This capability effectively monitors crop changes over time at the regional and state levels. Satellite-based applications often rely on widely recognized sensors such as MODIS, Sentinel-2 MSI, and PlanetScope [8, 46, 55]. These sensors offer spectral resolutions ranging from 4 to 13 bands, covering the electromagnetic spectrum's VIS, red-edge, and NIR regions. Their spatial resolution varies from 10 meters (Sentinel-2) to 500 meters (MODIS), making them well-suited for large-scale analysis, despite their limitations in detecting within-field variability. The availability of diverse spectral bands has facilitated calculating VI, such as NDVI and EVI, ranging from 1 to 12 indices across different studies. These indices are instrumental in evaluating plant health, water stress, and vegetative density.

The UAVs are the second most commonly used remote sensing platform for soybean yield prediction, as reported in **Table 2**. They were utilized in one-fifth of studies, accounting for approximately 38%. UAVs are particularly favored for their capacity to perform precise, low-altitude flights, enabling the capture of

| Reference | Remote sensing platform | Sensor type | Sensor model/manufacturer | Spectral resolution | Spatial resolution | Indices calculated |
|---|---|---|---|---|---|---|
| [33] | UAV | Multispectral | DJI Phantom 4 | 5 bands | 4 cm/pixel | 14 |
| [10] | UAV | Multispectral | DJI Matrice 600, DJI Phantom 4 | 5 bands | 3 cm/pixel | 33 |
| [42] | Satellite | Multispectral | MODIS | 7 bands | 250–500 m/pixel | 4 |
| [48] | Satellite | Multispectral | MODIS | 7 bands | 250–500 m/pixel | 7 |
| [31] | UAV | Hyperspectral | Cubert UHD 185 | 25 bands | 3 cm/pixel | 4 |
| [22] | Satellite | Multispectral | MODIS | 7 bands | 250–500 m/pixel | 9 |
| [9] | Ground-based | RGB | Hikrobot MV-CH250-90GC | 3 bands | 0.043 cm/pixel | — |
| [40] | Satellite | Multispectral | MODIS | 7 bands | 250–500 m/pixel | 5 |
| [17] | Satellite | Multispectral | Sentinel-2 MSI | 13 bands | 10 m/pixel | 12 |
| [5] | Satellite | Multispectral | MODIS | 7 bands | 250–500 m/pixel | 7 |
| [30] | UAV | Multispectral | Parrot Sequoia | 4 bands | 2.6–12 cm/pixel | 20 |
| [24] | Satellite | Multispectral | MODIS | 16 bands | 250–500 m/pixel | 4 |
| [4] | Satellite | Multispectral | MODIS | 16 bands | 250–500 m/pixel | 1 |
| [32] | UAV | Multispectral | DJI Phantom 4 | 5 bands | 3 cm/pixel | 1 |
| [49] | Satellite | Multispectral | MODIS | 7 bands | 250–500 m/pixel | 2 |
| [36] | Satellite | Multispectral | PlanetScope | 4 bands | 3.12 m/pixel | 11 |
| [39] | UAV | Multispectral | DJI Phantom 4 Pro | 5 bands | 1.6–2.7 cm/pixel | 3 |
| [28] | Satellite | Multispectral | Sentinel-2 MSI | 9 bands | 10 m/pixel | 8 |
| [15] | Satellite | Multispectral | Sentinel-2 MSI, Landsat 8 OLI | 13 bands | 10–30 m/pixel | 1 |
| [19] | Satellite | Multispectral | Sentinel-1 SAR, Sentinel-2 MSI | 13 bands | 10–20 m/pixel | 7 |
| [50] | UAV | Multispectral | Xacti CX-GB310 | 5 bands | 1 cm/pixel | 5 |
| [6] | UAV | Hyperspectral | UX4 Nuvem Rikola FPI2015 | 25 bands | 10 cm/pixel | 6 |
| [25] | UAV | RGB | DJI Phantom 4 Pro | 3 bands | 3.4 cm/pixel | — |

| Reference | Remote sensing platform | Sensor type | Sensor model/manufacturer | Spectral resolution | Spatial resolution | Indices calculated |
|---|---|---|---|---|---|---|
| [29] | UAV | RGB, Multispectral | DJI Phantom 4 RTK, Parrot Sequoia | 5 bands | 5 cm/pixel | 33 |
| [51] | UAV | Multispectral | DJI Phantom 4 RTK Parrot Sequoia | 5 bands | 0.75–2.8 cm/pixel | 6 |
| [52] | UAV | RGB | DJI Inspire 1 DJI Zenmuse X3 | 3 bands | 5–7 cm/pixel | 1 |
| [27] | UAV | Hyperspectral | DJI S1000 Cubert UHD 185 | 125 bands | 3 cm/pixel | 9 |
| [43] | Satellite | Multispectral | Sentinel-1 SAR, Sentinel-2 MSI | 10 bands | 10–20 m/pixel | 7 |
| [53] | Satellite | Multispectral | Sentinel-2 MSI | 10 bands | 10–20 m/pixel | 1 |
| [7] | Satellite | Multispectral | Sentinel-2 MSI, Landsat 8 OLI | 8 bands | 3–30 m/pixel | 4 |
| [54] | UAV | Hyperspectral | DJI Matrice 600 Pro Cubert UHD 185 | 125 bands | 3 cm/pixel | 9 |
| [55] | Satellite | Multispectral | MODIS | 7 bands | 250–500 m/pixel | 1 |
| [44] | Satellite | Multispectral | Sentinel-2 MSI, | 10 bands | 10–20 m/pixel | 4 |
| [56] | Satellite | Multispectral | Sentinel-1 SAR, Sentinel-2 MSI | 10 bands | 10–20 m/pixel | 7 |
| [13] | Ground-based | Hyperspectral | FieldSpec 3 Jr | 2151 bands | — | 4 |
| [57] | Satellite | Multispectral | Sentinel-2 MSI | 10 bands | 10–20 m/pixel | 1 |
| [8] | Satellite | Multispectral | Sentinel-2 MSI, | 10 bands | 3–10 m/pixel | 5 |
| [58] | Satellite | Multispectral | MODIS | 7 bands | 250–500 m/pixel | 2 |
| [37] | Satellite | Multispectral | MODIS | 7 bands | 250–500 m/pixel | 3 |
| [11] | Ground-based | Hyperspectral | FieldSpec 3 Jr | 2151 bands | 0.03–0.15 mm/pixel | 4 |
| [59] | UAV | Multispectral | SenseFly eBee RTK | 4 bands | 10 cm/pixel | 34 |
| [60] | Ground-based | Hyperspectral | FieldSpec 3 Jr | 2151 bands | — | 5 |
| [38] | UAV | RGB, Multispectral | DJI M300 SONY, MS600 V2 | 6 bands | 0.5–2 cm/pixel | 4 |
| [20] | UAV | Multispectral | DJI Phantom 4 RTK, Parrot Sequoia+ | 4 bands | 2.4–6.5 cm/pixel | 29 |

| Reference | Remote sensing platform | Sensor type | Sensor model/manufacturer | Spectral resolution | Spatial resolution | Indices calculated |
|---|---|---|---|---|---|---|
| [45] | UAV | RGB, Multispectral | DJI Phantom 4 RTK, Parrot Sequoia+ | 4 bands | 2–5 cm/pixel | 10 |
| [26] | Ground-based | Hyperspectral | UniSpec-DC | 250 bands | — | 5 |
| [41] | Satellite | Multispectral | Sentinel-2 MSI | 10 bands | 20 m/pixel | 1 |
| [61] | Satellite | Multispectral | PlanetScope MSI | 4 bands | 3 m/pixel | 1 |

**Table 2.**
*Summary of recent studies on soybean yield prediction using remote sensing techniques.*

high-resolution data that reveals fine-scale variations, such as changes in small leaves and canopy structure [6, 25]. This makes them highly effective for collecting accurate and timely field-level data. Among the most widely adopted UAV-mounted sensors are the DJI Phantom 4 Multispectral, MicaSense RedEdge-M, and Parrot Sequoia+, which typically provide 4–6 spectral bands, including RGB, red-edge, and NIR [10, 45]. These spectral capabilities are sufficient for calculating numerous VIs with high precision. With a spatial resolution between 2 and 5 cm per pixel, UAVs allow researchers to detect and monitor subtle differences in plant growth within individual fields. Consequently, several studies have demonstrated that UAVs with appropriate sensors can facilitate more than 30 VI computations. This wealth of spectral information improves the accuracy of yield prediction models and enhances the effectiveness of AI-based analytical techniques.

The least utilized remote sensing platform for soybean yield prediction is ground-based sensing, as indicated in **Table 2**, with only five studies, approximately 10%, reporting its use. Despite its limited adoption, ground-based platforms play a crucial role in research requiring high precision, particularly for detailed monitoring of plant physiological traits at the level of individual plants or within small experimental plots. These platforms are typically equipped with advanced spectroradiometers such as the FieldSpec 3 Jr. and HySpec VNIR/SWIR [13]. These instruments can accurately capture spectral reflectance across a broad wavelength range (350–2500 nm), offering more than 450 continuous bands with spectral resolutions between 3 and 5 nm. Such high spectral fidelity enables in-depth analysis of plant biochemical and structural attributes, including nitrogen content, chlorophyll concentration, water status, lignin presence, and tissue properties. Although ground-based sensors provide only point-based or near-point spatial coverage, they are essential for supporting field-level studies and calibrating airborne and satellite-based remote sensing models. Due to their high cost and limited coverage area, ground-based platforms are typically reserved for studies that require deep physiological analysis and spectral validation rather than large-scale monitoring.

## 4. Models for predicting soybean yield

A number of studies have explored the use of machine and deep learning models for predicting soybean yield based on remote sensing data. As shown in **Figure 4**, approximately 74% of the reviewed studies employed machine learning approaches, while 26% utilized deep learning techniques. The selection between machine learning and deep learning models largely depends on factors such as the study methodology, the algorithm's complexity, data requirements, practical applications, and the specific challenges addressed. Machine learning methods, including RF, SVM, and Artificial Neural Networks (ANN), are widely used for their relatively simple architecture, faster training times, and ability to work effectively with smaller, structured datasets [24, 38, 39]. These models typically require manual feature extraction and selection, for example, using VI derived from remote sensing imagery as input features. In contrast, deep learning methods, such as CNN, LSTM, and Deep Neural Networks (DNN), can automatically extract features from raw data, such as high-dimensional remote sensing images [48, 59, 62]. This advantage reduces the need for manual preprocessing and allows for improved spatial and temporal data handling, such as time-series imagery captured throughout the growing season. However, deep learning models often require large datasets for training and demand significant

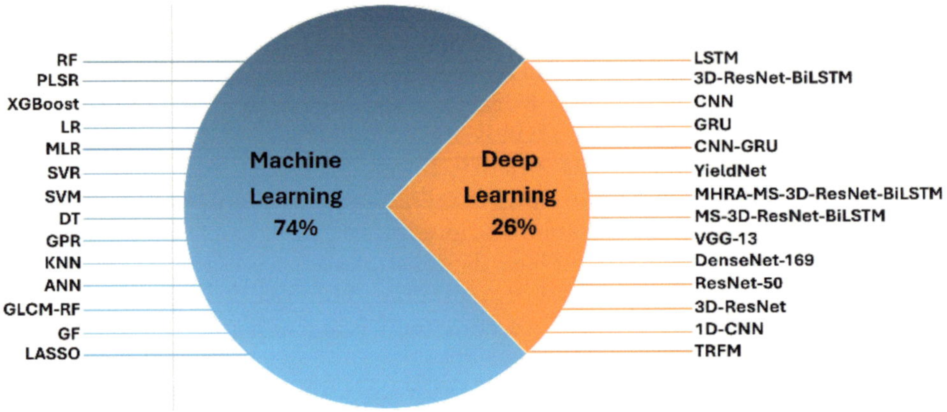


**Figure 4.**
*Proportion of machine learning and deep learning models used in soybean yield prediction studies.*

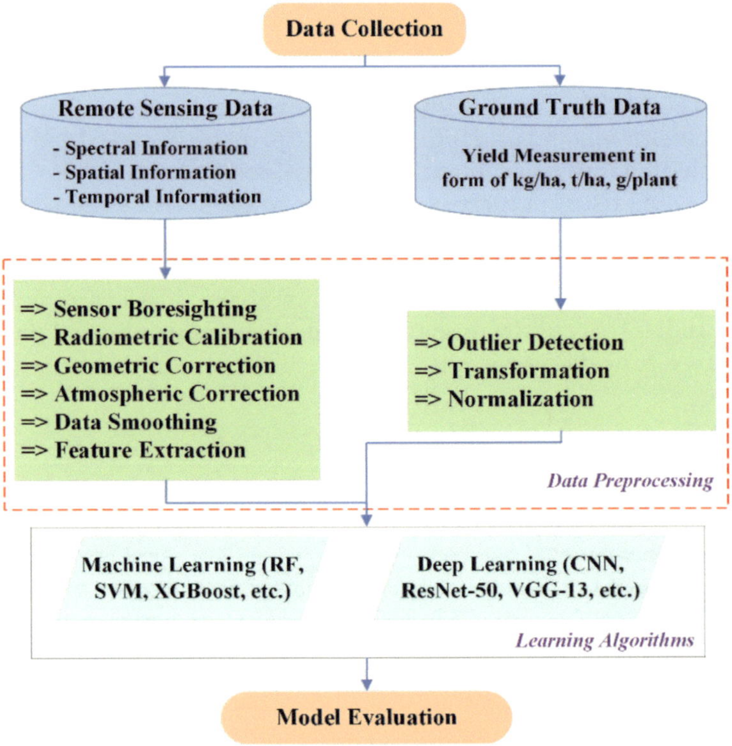


**Figure 5.**
*Flowchart for developing models for soybean yield prediction using remote sensing.*

computational resources. While machine learning models may struggle with high-dimensional data without proper feature engineering, deep learning models are well-suited for image-rich and unstructured datasets, offering enhanced performance in high data volume and complexity. Therefore, selecting the appropriate model type is crucial for maximizing predictive accuracy in soybean yield modeling using remote sensing data.

**Figure 5** illustrates a comprehensive workflow for soybean yield prediction that integrates remote sensing imagery with ground truth yield data. The process begins with data collection, which involves acquiring two primary datasets: remote sensing data, providing spectral, spatial, and temporal information captured from various platforms (e.g., satellites, UAVs, or ground-based sensors), and ground truth data, representing actual yield measurements typically expressed in units such as kg/ha, t/ha, or g/plant, depending on the study design. Following data acquisition, a critical step is data preprocessing, where each data type undergoes tailored processing routines. For remote sensing data, this includes sensor boresighting (particularly relevant for airborne platforms), radiometric calibration, geometric and atmospheric corrections, data smoothing, and feature extraction, often involving the derivation of VI or textural features. Meanwhile, ground truth data is subjected to outlier detection, data transformation, and normalization to ensure consistency with remotely sensed inputs. Once the data is preprocessed, predictive modeling is performed using machine or deep learning algorithms. These models are calibrated using training data and validated against independent test datasets. Model performance is typically evaluated using statistical indicators, with the coefficient of determination ($R^2$) serving as a common metric to assess the goodness of fit between predicted and observed yield values.

## 5. Challenges and future direction

Despite significant progress in leveraging remote sensing for soybean yield prediction, several technical and operational challenges remain. One of the most pressing issues is the scarcity of high-quality and consistent ground truth data, which complicates model calibration and validation efforts. In addition, model complexity, often influenced by heterogeneous factors such as weather, soil properties, and crop management, limits the generalizability of predictive models across regions and seasons. The temporal and spatial resolution of imagery poses another constraint, as remote sensing data must be acquired during critical phenological stages to ensure accurate yield estimation, which is not always feasible due to cloud cover or acquisition scheduling limitations. Moreover, integrating multisource data, including multispectral, hyperspectral, and thermal imagery, introduces challenges in data synchronization, requiring advanced fusion techniques and computational resources. Phenological variability further complicates modeling efforts, as variations in soybean development stages across different environments are not always well represented in current algorithms.

To address these limitations, future research should focus on advancing machine learning techniques, particularly by integrating deep learning architectures and transfer learning to overcome data limitations. Multimodal data fusion, combining spectral data from diverse platforms, is essential to improve model robustness and prediction accuracy. Additionally, real-time monitoring and adaptive systems should be developed to provide timely interventions during the growing season, supported by innovations in the Internet of Things (IoT) and sensor networks. Improving phenological modeling by identifying optimal imaging windows and correlating growth stages with yield dynamics can also enhance predictive power. Lastly, collaborative frameworks between researchers, growers, and agronomists, facilitated by open data sharing and best-practice dissemination, will ensure that remote sensing technologies transition from experimental tools to scalable solutions that benefit farmers and promote sustainable soybean production.

## 6. Conclusion

This chapter presents a comprehensive review of the application of remote sensing technologies for soybean yield prediction, emphasizing current capabilities, limitations, and future directions. Four major remote sensing platforms were analyzed: spaceborne, airborne, UAV-based, and ground-based. Among the reviewed studies, satellite (spaceborne) platforms were the most used due to their ability to cover large agricultural areas with consistent temporal resolution throughout the growing season. However, their performance is often constrained by adverse weather conditions such as cloud cover, wind, and snow, and the high cost of high-resolution imagery. Interestingly, no study was found to have applied airborne platforms for soybean yield prediction. This absence may be attributed to the high operational costs and the technical complexity associated with deploying and managing such systems, particularly when compared to the increasing accessibility of UAVs. In contrast, UAV and ground-based platforms, applied in 36 and 10% of the studies, offer higher spatial resolution and more flexible deployment. These platforms are particularly valuable for detailed monitoring at the field or plant level, despite limitations in coverage area and operational complexity. Regarding modeling approaches, machine learning algorithms were employed in 74% of the studies, while 26% utilized deep learning techniques. Combined with spectral imagery, these models have demonstrated significant potential for predicting soybean yield by capturing critical biophysical and biochemical indicators throughout the crop cycle. Future research should focus on enhancing remote sensing-based technologies' accuracy, robustness, and operational simplicity to provide growers with rapid, real-time guidance. Integrating improved AI models with accessible decision-support systems will be key to maximizing practical impact. Additionally, greater collaboration with producers and expansion of on-farm trials will be essential to validate remote sensing applications under real-world conditions and promote broader adoption in soybean production.

## Acknowledgements

The author gratefully acknowledges the University of Maryland Eastern Shore (UMES), Department of Agriculture, Food, and Resource Sciences, for their support and resources that contributed to the development of this chapter.

## Nomenclature

| | |
|---|---|
| VIS | visible |
| NIR | near-infrared |
| SWIR | shortwave infrared |
| VI | vegetation index |
| AI | artificial intelligence |
| UAV | unmanned aerial vehicle |
| RGB | red-green-blue |
| LiDAR | light detection and ranging |
| RF | random forest |
| SVM | support vector machine |
| kNN | k-nearest neighbors |

| | |
|---|---|
| CNN | convolutional neural network |
| LSTM | long short-term memory |
| NDVI | normalized difference vegetation index |
| EVI | enhanced vegetation index |
| NDWI | normalized difference water index |
| PRI | photochemical reflectance index |
| MODIS | moderate resolution imaging spectroradiometer |
| MSI | MultiSpectral instrument |
| ANN | artificial neural network |
| DNN | deep neural network |
| 3D-ResNet-BiLSTM | 3D residual network with bi-directional LSTM |
| GRU | gated recurrent unit |
| CNN-GRU | convolutional neural network with gated recurrent unit |
| YieldNet | yield prediction network |
| MHRA-MS-3D-ResNet-BiLSTM | multi-head residual attention with multi-stream 3D-ResNet and BiLSTM |
| MS-3D-ResNet-BiLSTM | multi-stream 3D residual network with BiLSTM |
| VGG-13 | visual geometry group network with 13 layers |
| DenseNet-169 | densely connected convolutional network with 169 layers |
| ResNet-50 | 50-layer residual neural network |
| 3D-ResNet | three-dimensional residual network |
| 1D-CNN | one-dimensional convolutional neural network |
| TRFM | transformer |
| PLSR | partial least squares regression |
| XGBoost | extreme gradient boosting |
| LR | linear regression |
| MLR | multiple linear regression |
| SVR | support vector regression |
| DT | decision tree |
| GPR | Gaussian Process Regression |
| GLCM-RF | gray-level co-occurrence matrix – random forest |
| GM | geometric features |
| LASSO | least absolute shrinkage and selection operator |
| $R^2$ | coefficient of determination |
| IoT | Internet of Things |

## Author details

Alfadhl Y. Alkhaled
Department of Agriculture, Food, and Resource Sciences, University of Maryland Eastern Shore, Princess Anne, MD, USA

*Address all correspondence to: alkhaled@umes.edu

## References

[1] Nations F and AO of the U. FAOSTAT: Crops and livestock products. 2024. Available from: https://www.fao.org/faostat/en/#data/QCL

[2] United States Department of Agriculture. Oilseeds: World Markets and Trade. Washington, DC, USA: Foreign Agricultural Service; 2023. Available from: https://www.fas.usda.gov/data/oilseeds-world-markets-and-trade

[3] U.S. Energy Information Administration. Biodiesel Explained: Use of Biodiesel. Washington, D.C., USA: U.S. Energy Information Administration; 2024

[4] Xie W, Huang Y, Meng Q. Soybean yield Modeling and analysis with weather dynamics in the greater Mississippi River basin. Climate. 2025;**13**:1-18. DOI: 10.3390/cli13020033

[5] Li Y, Zeng H, Zhang M, et al. Global de-trending significantly improves the accuracy of XGBoost-based county-level maize and soybean yield prediction in the Midwestern United States. GIScience & Remote Sensing. 2024;**61**:1-26. DOI: 10.1080/15481603.2024.2349341

[6] Berveglieri A, Imai NN, Watanabe FSY, et al. Remote prediction of soybean yield using UAV-based hyperspectral imaging and machine learning models. AgriEngineering. 2024;**6**:3242-3260

[7] Amankulova K, Farmonov N, Akramova P, et al. Comparison of PlanetScope, Sentinel-2, and landsat 8 data in soybean yield estimation within-field variability with random forest regression. Heliyon. 2023;**9**:1-17. DOI: 10.1016/j.heliyon.2023.e17432

[8] Amankulova K, Farmonov N, Omonov K, et al. Integrating the Sentinel-1, Sentinel-2 and topographic data into soybean yield modelling using machine learning. Advances in Space Research. 2024;**73**:4052-4066

[9] Li X, Chen M, He S, et al. Estimation of soybean yield based on high-throughput phenotyping and machine learning. Frontiers in Plant Science. 2024;**15**:1-17

[10] Skobalski J, Sagan V, Alifu H, et al. Bridging the gap between crop breeding and GeoAI: Soybean yield prediction from multispectral UAV images with transfer learning. ISPRS Journal of Photogrammetry and Remote Sensing. 2024;**210**:260-281

[11] Giri A, Sagan V, Alifu H, et al. A wavelet decomposition method for estimating soybean seed composition with hyperspectral data. Remote Sensing. 2024;**16**:1-17

[12] Archontoulis SV, Castellano MJ, Licht MA, et al. Predicting crop yields and soil-plant nitrogen dynamics in the US corn belt. Crop Science. 2020;**60**:721-738

[13] Crusiol LGT, Nanni MR, Furlanetto RH, et al. Yield prediction in soybean crop grown under different levels of water availability using reflectance spectroscopy and partial least squares regression. Remote Sensing. 2021;**13**:1-21

[14] Yuan J, Zhang Y, Zheng Z, et al. Grain crop yield prediction using machine learning based on UAV remote sensing: A systematic literature review. Drones. 2024;**8**:1-27. DOI: 10.3390/drones8100559

[15] Andrade TG, Junior ASDA, Souza MO, et al. Soybean yield prediction using remote sensing in

Southwestern Piauí state, Brazil. Revista Caatinga. 2022;**35**:105-116

[16] Wang F, Li J, Peng D, et al. Estimating soybean yields using causal inference and deep learning approaches with satellite remote sensing data. IEEE Journal of Selected Topics in Applied Earth Observations and Remote Sensing. 2024;**17**:14161-14178

[17] Amaral LR, Oldoni H, Baptista GMM, et al. Remote sensing imagery to predict soybean yield: A case study of vegetation indices contribution. Precision Agriculture. 2024;**25**:2375-2393

[18] Vidican R, Mălinaș A, Ranta O, et al. Using remote sensing vegetation indices for the discrimination and monitoring of agricultural crops: A critical review. Agronomy. 2023;**13**:1-27

[19] Fathi M, Shah-Hosseini R, Moghimi A, et al. MHRA-MS-3D-ResNet-BiLSTM: A multi-head-residual attention-based multi-stream deep learning model for soybean yield prediction in the U.S. using multi-source remote sensing data. Remote Sensing. 2025;**17**:1-25. DOI: 10.3390/rs17010107

[20] Habibi LN, Matsui T, Tanaka TST. Critical evaluation of the effects of a cross-validation strategy and machine learning optimization on the prediction accuracy and transferability of a soybean yield prediction model using UAV-based remote sensing. Journal of Agriculture and Food Research. 2024;**16**:101096

[21] Alkhaled AY, Wang Y. Developing a robust yield prediction model for potatoes (*Solanum tuberosum* L.) using multi-faceted and multi-year data. Smart Agricultural Technology. 2025;**10**:1-12. DOI: 10.1016/j.atech.2024.100734

[22] Li J, Zhang W, Zhou H, et al. Weed detection in soybean fields using improved YOLOv7 and evaluating herbicide reduction efficacy. Frontiers in Plant Science. 2023;**14**:1-17. DOI: 10.3389/fpls.2023.1284338

[23] Negrisoli MM, Negrisoli RM, da Silva FN, et al. Soybean rust detection and disease severity classification by remote sensing. Agronomy Journal. 2022;**114**:3246-3262

[24] von Bloh M, de Nóia Júnior RS, Wangerpohl X, et al. Machine learning for soybean yield forecasting in Brazil. Agricultural and Forest Meteorology. 2023;**341**:1-12. DOI: 10.1016/j.agrformet.2023.109670

[25] Bhadra S, Sagan V, Skobalski J, et al. End-to-end 3D CNN for plot-scale soybean yield prediction using multitemporal UAV-based RGB images. Precision Agriculture. 2024;**25**:834-864

[26] Yoosefzadeh-Najafabadi M, Tulpan D, Eskandari M. Using hybrid artificial intelligence and evolutionary optimization algorithms for estimating soybean yield and fresh biomass using hyperspectral vegetation indices. Remote Sensing. 2021;**13**:1-21. DOI: 10.3390/rs13132555

[27] Wang L, Gao R, Li C, et al. Mapping soybean maturity and biochemical traits using UAV-based hyperspectral images. Remote Sensing. 2023;**15**:1-20

[28] Crusiol LGT, Sun L, Sibaldelli RNR, et al. Strategies for monitoring within-field soybean yield using Sentinel-2 Vis-NIR-SWIR spectral bands and machine learning regression methods. Precision Agriculture. 2022;**23**:1093-1123

[29] Sun G, Zhang Y, Chen H, et al. Improving soybean yield prediction by integrating UAV nadir and cross-circling oblique imaging. European Journal of Agronomy. 2024;**155**:127134

[30] Alabi TR, Abebe AT, Chigeza G, et al. Estimation of soybean grain yield from multispectral high-resolution UAV data with machine learning models in West Africa. Remote Sensing Applications: Society and Environment. 2022;**27**:100782

[31] Ren P, Li H, Han S, et al. Estimation of soybean yield by combining maturity group information and unmanned aerial vehicle multi-sensor data using machine learning. Remote Sensing. 2023;**15**:1-23. DOI: 10.3390/rs15174286

[32] Maulit A, Nugumanova A, Apayev K, et al. A multispectral UAV imagery dataset of wheat, soybean and barley crops in East Kazakhstan. Data (Basel). 2023;**8**:1-13. DOI: 10.3390/data8050088

[33] Shammi SA, Huang Y, Feng G, et al. Application of UAV multispectral imaging to monitor soybean growth with yield prediction through machine learning. Agronomy. 2024;**14**:1-18

[34] Zhou Z, Majeed Y, Diverres Naranjo G, et al. Assessment for crop water stress with infrared thermal imagery in precision agriculture: A review and future prospects for deep learning applications. Computers and Electronics in Agriculture. 2021;**182**:1-10. DOI: 10.1016/j.compag.2021.106019

[35] Farhan SM, Yin J, Chen Z, et al. A comprehensive review of LiDAR applications in crop management for precision agriculture. Sensors. 2024;**24**:1-44. DOI: 10.3390/s24165409

[36] Joshi DR, Clay SA, Sharma P, et al. Artificial intelligence and satellite-based remote sensing can be used to predict soybean (Glycine max) yield. Agronomy Journal. 2024;**116**:917-930

[37] Farmonov N, Amankulova K, Khan SN, et al. Effectiveness of machine learning and deep learning models at county-level soybean yield forecasting. Hungarian Geographical Bulletin. 2023;**72**:383-398

[38] Da H, Li Y, Xu L, et al. Advancing soybean biomass estimation through multi-source UAV data fusion and machine learning algorithms. Smart Agricultural Technology. 2025;**10**:1-16. DOI: 10.1016/j.atech.2025.100778

[39] Pérez O, Diers B, Martin N. Maturity prediction in soybean breeding using aerial images and the random Forest machine learning algorithm. Remote Sensing. 2024;**16**:1-27. DOI: 10.3390/rs16234343

[40] Khaki S, Pham H, Wang L. Simultaneous corn and soybean yield prediction from remote sensing data using deep transfer learning. Scientific Reports. 2021;**11**:1-14

[41] Gaso DV, Cue La Rosa LE, Puntel LA, et al. Predicting spatial and temporal variability in soybean yield using deep learning and open source data. European Journal of Agronomy. 2025;**164**:127498

[42] Wang Y, Feng K, Sun L, et al. Satellite-based soybean yield prediction in Argentina: A comparison between panel regression and deep learning methods. Computers and Electronics in Agriculture. 2024;**221**:108978

[43] Fathi M, Shah-Hosseini R, Moghimi A. 3D-ResNet-BiLSTM model: A deep learning model for county-level soybean yield prediction with time-series Sentinel-1, Sentinel-2 imagery, and Daymet data. Remote Sensing. 2023;**15**:1-20

[44] Sarkar S, Sagan V, Bhadra S, et al. Spectral enhancement of PlanetScope using Sentinal-2 images to estimate

soybean yield and seed composition. Scientific Reports. 2024;**14**:1-16

[45] dos Lima CS, Junior DFU, Carvalho IR, et al. Use of Phenomics in the selection of UAV-based vegetation indices and prediction of agronomic traits in soybean subjected to flooding. AgriEngineering. 2024;**6**:3261-3278

[46] Amankulova K, Farmonov N, Abdelsamei E, et al. A novel fusion method for soybean yield prediction using sentinel-2 and PlanetScope imagery. IEEE Journal of Selected Topics in Applied Earth Observations and Remote Sensing. 2024;**17**:13694-13707

[47] Curran PJ. Remote sensing of foliar chemistry. Remote Sensing of Environment. 1989;**30**:271-278

[48] Lu J, Fu H, Tang X, et al. GOA-optimized deep learning for soybean yield estimation using multi-source remote sensing data. Scientific Reports. 2024;**14**:1-19

[49] Batistella D, Modolo AJ, da Campos JRR, et al. Comparative analysis of orbital sensors in soybean yield estimation by the random forest algorithm. Ciencia e Agrotecnologia. 2023;**47**:1-9. DOI: 10.1590/1413-7054202347002423

[50] Sakurai K, Toda Y, Kajiya-Kanegae H, et al. Time-series multispectral imaging in soybean for improving biomass and genomic prediction accuracy. Plant Genome. 2022;**15**:1-18

[51] Zhou L, Zhang Y, Chen H, et al. Soybean yield estimation and lodging classification based on UAV multi-source data and self-supervised contrastive learning. Computers and Electronics in Agriculture. 2025;**230**:109822

[52] Toda Y, Kaga A, Kajiya-Kanegae H, et al. Genomic prediction modeling of soybean biomass using UAV-based remote sensing and longitudinal model parameters. Plant Genome. 2021;**14**:1-13

[53] de Freitas RG, Oldoni H, Joaquim LF, et al. Predicting on-farm soybean yield variability using texture measures on Sentinel-2 image. Precision Agriculture. 2024:2977-3000

[54] Sun G, Zhang Y, Wang L, et al. Bridging the gap between hyperspectral imaging and crop breeding: Soybean yield prediction and lodging classification with prototype contrastive learning. Computers and Electronics in Agriculture. 2025;**230**:109859

[55] Stepanov A, Dubrovin K, Sorokin A. Function fitting for modeling seasonal normalized difference vegetation index time series and early forecasting of soybean yield. Crop Journal. 2022;**10**:1452-1459

[56] Gaso DV, Paudel D, de Wit A, et al. Beyond assimilation of leaf area index: Leveraging additional spectral information using machine learning for site-specific soybean yield prediction. Agricultural and Forest Meteorology. 2024;**351**:110022

[57] Gaso DV, de Wit A, Berger AG, et al. Predicting within-field soybean yield variability by coupling Sentinel-2 leaf area index with a crop growth model. Agricultural and Forest Meteorology. 2021;**308-309**:108553

[58] Maas MD, Salvia M, Spennemann PC, et al. Robust multisensor prediction of drought-induced yield anomalies of soybeans in Argentina. IEEE Geoscience and Remote Sensing Letters. 2022;**19**:1-4

[59] Teodoro PE, Teodoro LPR, Baio FHR, et al. Predicting days to maturity, plant height, and grain yield in soybean: A machine and deep learning approach

using multispectral data. Remote Sensing. 2021;**13**:1-15. DOI: 10.3390/rs13224632

[60] Crusiol LGT, Nanni MR, Sibaldelli RNR, et al. Early modeling of the upcoming landsat next constellation for soybean yield prediction under varying levels of water availability. Remote Sensing. 2024;**16**:1-29. DOI: 10.3390/rs16224184

[61] Fischer RJ, Rekabdarkolaee HM, Joshi DR, et al. Soybean prediction using computationally efficient Bayesian spatial regression models and satellite imagery. Agronomy Journal. 2024;**116**:2841-2849

[62] Darwin B, Dharmaraj P, Prince S, et al. Recognition of bloom/yield in crop images using deep learning models for smart agriculture: A review. Agronomy. 2021;**11**:1-22

Chapter 6

# Advanced Machine Learning Approaches for Soybean Yield Prediction in Uttarakhand

*Yunish Khan, Vinod Kumar, Parul Setiya and Anurag Satpathi*

## Abstract

Accurate pre-harvest forecasting of soybean yields is essential for developing, implementing, and refining food security policies, as well as for optimizing agricultural storage and marketing systems. Since soybean growth and development are inherently influenced by climatic conditions, forecasting models that incorporate weather variables are vital for generating reliable yield predictions. However, selecting the most appropriate prediction model for soybean production remains a complex challenge. This research developed three multivariate models to estimate soybean yields across eight key districts of Uttarakhand: Udham Singh Nagar, Almora, Uttarkashi, Dehradun, Pauri-Garhwal, Tehri-Garhwal, Rudraprayag, and Pithoragarh. The models employed were artificial neural networks (ANN), principal component analysis integrated with ANN (PCA-ANN), and the least absolute shrinkage and selection operator (LASSO). Historical time series data of soybean yields and corresponding weather indices were utilized to develop and validate the predictive models. The dataset was partitioned into training (80%) and validation (20%) subsets. Model development was based on average values calculated for distinct soybean phenological phases, derived from daily weather observations. Both weighted and unweighted forms of weather indices were analyzed to optimize model performance. Among the evaluated models, the PCA-ANN approach demonstrated the highest forecasting accuracy. The overall performance ranking across all studied regions was: PCA-ANN > ANN > LASSO. The PCA-ANN hybrid model emerged as the most robust and consistent method for estimating soybean yields in the selected districts of Uttarakhand, providing critical support for agricultural management and policy planning in soybean production systems.

**Keywords:** crop yield prediction, artificial neural network, principal component analysis, least absolute and selection operator, soybean production

## 1. Introduction

Agriculture serves as a cornerstone of both global and Indian economies. The ever-growing global population, coupled with the challenges posed by climate change, has increasingly threatened agricultural productivity. Although it is not entirely possible to eliminate these issues, having timely access to forward-looking insights allows farmers to formulate effective strategies and take precautionary measures. Early communication of crop yield forecasts can significantly contribute to minimizing the risks tied to food shortages. Accordingly, the importance of accurate yield prediction has become more pronounced than ever [1]. Trustworthy yield assessments are vital in the agricultural sector, as they offer essential knowledge about anticipated harvest levels, enabling swift and informed decisions by both farmers and other key stakeholders [2]. Precise forecasting equips farmers to allocate resources more wisely, reduce inefficiencies, and improve overall productivity [3]. Furthermore, it supports policymakers in making timely, informed decisions related to food security, grain reserves, transportation logistics, market planning, and price regulation [4, 5].

Globally, soybean (*Glycine max (L.) Merrill*) stands out as one of the most essential oilseed crops cultivated worldwide [6]. Brazil leads in soybean production, contributing 38% of global output, followed by the United States at 31%. India holds the fifth position among the top soybean-producing nations [7]. In India's northwestern Himalayan hill zone, soybean is predominantly cultivated as a major Kharif season crop. In this region, Uttarakhand holds a significant position, contributing nearly 90–95% of the total land dedicated to soybean cultivation and overall yield [8].

Phenological weather indices have become essential components in agricultural research focused on yield estimation, as they indicate the occurrence of crucial crop development stages—such as flowering and ripening—based on environmental parameters like temperature, rainfall, and solar radiation [9–13]. These indices shed light on a crop's physiological progression, which is closely associated with its productivity [14]. In earlier practices, yield estimation primarily depended on labor-intensive crop-cutting surveys and traditional statistical techniques such as Multivariate Linear Regression (MLR), which often lacked sufficient accuracy [15, 16]. Over the past few years, machine learning algorithms have become increasingly popular due to their ability to manage vast and complex datasets, uncover hidden patterns, and generate more precise forecasts. These sophisticated models have been successfully utilized to predict yields of several crops, including rice [5], wheat [3, 17], pigeon pea [18], cashew [19], sorghum [20], and coconut [21].

Previous studies have largely compared the effects of direct weather parameters and derived weather indices on crop yield prediction. However, only a few have investigated the potential of integrated or hybrid models and the interplay between individual predictive techniques [22–24]. To bridge this gap, our research develops hybrid models—particularly a PCA-ANN (principal component analysis-artificial neural network) approach using weather indices. The primary goal of this investigation is to examine the accuracy and robustness of statistical and machine learning techniques for forecasting soybean production. Through this approach, we strive to identify the most accurate and efficient technique. The findings hold considerable relevance for the agricultural sector, offering critical insights into diverse yield prediction methodologies. By highlighting the advantages and limitations of each model, this research contributes to better soybean production planning, efficient resource allocation, and enhanced farm productivity. Additionally, the outcomes may support policymakers and stakeholders in making strategic decisions related to food security, crop management, and market planning. Ultimately, the study aspires to advance the science of yield forecasting and promote more dependable agricultural prediction tools [25, 26].

### 1.1 Research objectives

The primary objective of this investigation is to examine the accuracy and robustness of statistical and machine learning techniques for forecasting soybean production, striving to identify the most accurate and efficient technique. The findings hold considerable relevance for the agricultural sector, offering critical insights into diverse yield prediction methodologies. By highlighting the advantages and limitations of each model, this research contributes to better soybean production planning, efficient resource allocation, and enhanced farm productivity. Additionally, the outcomes may support policymakers and stakeholders in making strategic decisions related to food security, crop management, and market planning. Ultimately, this study aspires to advance the science of yield forecasting and promote more dependable agricultural prediction tools.

## 2. Materials and methods

### 2.1 Study location

This research formulated yield prediction models using Kharif season soybean production data alongside long-term meteorological records. The study focused on eight major soybean-growing districts in Uttarakhand: Udham Singh Nagar, Almora, Uttarkashi, Dehradun, Pauri-Garhwal, Tehri-Garhwal, Rudraprayag, and Pithoragarh. These regions are characterized by nutrient-rich soils and excellent water-holding capacity, which create optimal conditions for soybean growth and result in consistently abundant yields. The selected districts represent diverse agro-climatic zones within Uttarakhand, ensuring that the developed models capture the variability in growing conditions across the state. This geographic diversity enhances the robustness and applicability of the yield prediction models.

### 2.2 Data collection and sources

Time series data on soybean yield covering a 20-year period (2001–2020) were sourced from multiple reliable institutions. For the Udham Singh Nagar district, yield data were obtained from the Soybean Breeding Laboratory at G.B. Pant University of Agriculture and Technology (GBPUAT), Pantnagar. For the remaining districts-Uttarkashi, Dehradun, Pauri-Garhwal, Tehri-Garhwal, Rudraprayag, and Pithoragarh-data were acquired from the Directorate of Economics and Statistics, Ministry of Agriculture and Farmers Welfare. Meteorological information, including daily maximum and minimum temperatures, average relative humidity, rainfall, sunshine duration, and wind velocity, was collected from complementary sources. For Udham Singh Nagar, weather data were obtained from the Department of Agrometeorology at GBPUAT, while data for the remaining districts were retrieved from the NASA POWER portal (https://power.larc.nasa.gov/data-access-viewer/). These weather parameters were subsequently converted into phenological stage-specific average values for further analysis. The soybean crop growth stages were categorized according to Standard Meteorological Weeks, providing a systematic framework for analyzing weather impacts during critical development periods.

## 2.3 Methodology

### *2.3.1 Development of weighted and unweighted weather indices*

This research utilized a comprehensive dataset covering a 20-year period. The dataset was partitioned into two subsets: the first, containing 16 years of information (80%), was applied for model training, while the remaining 4 years (20%) were designated for testing and validation purposes [27, 28]. For the phenological evaluation, daily climatic data were processed to obtain average values, which were then utilized to generate two classifications of agro-climatic indices—namely, weighted and unweighted [29]. For each weather parameter, both types of indices were created. The unweighted indices were derived by summing the values of individual weather variables or their pairwise interactions. Alternatively, the weighted indices were formulated by associating each variable or interaction with its respective correlation coefficient concerning the target variable and subsequently aggregating the resulting values. For the creation of these meteorological indices, all six climatic elements were combined in every possible pair, leading to 15 unique pairwise interactions and generating 30 indices based on these interactions. Moreover, 12 additional indices were calculated using individual weather parameters, excluding their interactions. The formulas suggested by Satpathi et al. [5] were followed for deriving both the weighted and unweighted indices, as indicated in Eqs. (1) and (2), respectively.

*Unweighted weather indices*:

$$Z_{ij} = \sum_{p=1}^{n} X_{ip},\ Z_{ij} = \sum_{p=1}^{n} X_{ip}\, X_{i'p} \tag{1}$$

*Weighted weather indices*:

$$Z_{ij} = \sum_{p=1}^{n} r_{ip}^{j}\, X_{ip},\ Z_{ii'j} = \sum_{p=1}^{n} r_{ii'p}^{j}\, X_{ip}\, X_{i'p} \tag{2}$$

where $n$ is the number of phenological stages. $X_{ip}/X_{i'p}$ is the value of $i^{th}/\ i'^{th}$ weather variable and $r_{ip}/r_{ii'p}$ is the value of the correlation coefficient of yield with $i^{th}$ weather variable/ product of $i^{th}$ and $i'^{th}$ weather variables during $p^{th}$ phenological stages. $j = 0$ represents unweighted weather indices and $j = 1$ represents weighted weather indices. By following the above procedure, a total of 42 weather indices were generated.

### *2.3.2 Model development framework*

The systematic progression undertaken for model formulation involved several key stages. Three multivariate models were developed utilizing historical records of soybean yield and phenological weather indices. The models were trained and tested using these datasets. The three models utilized were: ANN, PCA-ANN, and LASSO. The development process followed a structured approach:

1. Data preprocessing and quality control

2. Weather index calculation

3. Feature selection and dimensionality reduction

4. Model training and hyperparameter tuning

5. Model validation and performance evaluation

6. Comparative analysis and ranking.

#### *2.3.3 Multivariate techniques for soybean yield prediction model development*

The specifics of the multivariate analytical methods employed in this research for constructing the soybean yield forecasting model are outlined below:

##### *2.3.3.1 Principal component analysis (PCA)*

The primary purpose of principal component analysis (PCA) is to reduce the dimensionality of a dataset while retaining the majority of the original data's variance and information content. In soybean yield prediction, PCA is utilized to minimize the likelihood of overfitting caused by high dimensionality and strong correlations among weather-related independent variables. PCA is also recognized as a method for feature reduction, data compression, and dimensional complexity reduction. All input variables were standardized by subtracting the mean value and dividing by the standard deviation to ensure equal weighting across different scales of weather parameters. Only the principal components (PCs) with eigenvalues greater than 1 were retained for analysis, following the Kaiser criterion [4, 30].

In soybean yield studies, PCA has proven particularly effective in identifying the most influential weather patterns that affect crop development stages. For instance, temperature and precipitation variables during flowering and pod-filling stages often contribute significantly to the first few principal components, which typically explain 70–85% of the total variance in soybean yield prediction models.

##### *2.3.3.2 Artificial neural network (ANN)*

Artificial neural networks (ANNs) are computational models inspired by the neural structure and functioning of the human brain, primarily utilized in data-driven learning systems for complex pattern recognition tasks [31–33]. These models consist of interconnected processing units known as "neurons," which process input data and generate outputs by transmitting signals through multiple layers of the network [34]. An ANN typically consists of three primary components: an input layer, one or more hidden layers, and an output layer. Data flows in a forward direction from the input layer, through the hidden layer(s), and finally to the output layer [35]. The input layer features nodes representing both weighted and unweighted weather indices relevant to soybean growth, whereas the output layer contains the response variable (soybean yield). The hidden layer comprises neurons that perform mathematical transformations through activation functions, commonly using sigmoid, ReLU, or tanh functions. For soybean yield prediction, ANNs have demonstrated superior performance in capturing non-linear relationships between weather variables and yield outcomes. Studies have shown that ANNs can effectively model the complex interactions between temperature, precipitation, solar radiation, and humidity during critical growth stages such as emergence, flowering, and pod development. Neurons in one layer are connected to those in the subsequent layer *via* weighted connections, each carrying a specific weight value [31]. The final output of the ANN is computed by taking the weighted summation of the input features, which is then passed through an activation function to produce the network's prediction. The number of input nodes corresponds to the total number of explanatory variables utilized in the model [26, 36–38]. One of the major challenges in constructing ANN models for soybean yield prediction is selecting the appropriate number of neurons in the hidden layer. In the current study, this was resolved using the "train" function from the "caret" library in R, employing the "nnet" approach with 10-fold cross-validation [39]. Research has shown that for soybean yield prediction, optimal hidden layer sizes typically range from 3 to 15 neurons, depending on the number of input variables and dataset complexity.

To optimize the model, the "tuneGrid" function was utilized to determine the most suitable number of hidden units for improving forecasting accuracy. During the model development phase, the iteration count was limited to a maximum of 100 to prevent overfitting. Additionally, weight regularization was implemented through a hyperparameter called "decay." The decay parameter acts as a regularization mechanism to prevent overfitting by applying a penalty to overly complex models. This concept is similar to statistical shrinkage, where model complexity is controlled to enhance generalization capability. All calculated weather indices were used as input features, while soybean yield served as the response variable (**Figure 1**).

*2.3.3.3 Least absolute shrinkage and selection operator (LASSO)*

The LASSO approach serves two key functions: regularization and feature selection, making it an effective and versatile technique for soybean yield modeling. Through a shrinkage mechanism, LASSO imposes a penalty on the regression coefficients, causing some of them to shrink to zero. Consequently, only the predictor variables with non-zero coefficients are retained in the final model, effectively identifying the most important weather variables for soybean yield prediction [40]. In soybean yield studies, LASSO has proven particularly valuable for identifying critical weather windows and eliminating redundant variables. For example, LASSO

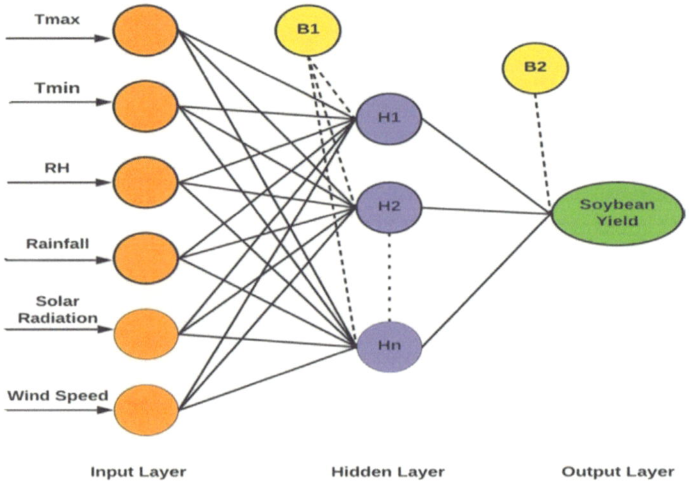


**Figure 1.**
*Flow diagram of the ANN model development for soybean yield prediction.*

often selects temperature during reproductive stages and precipitation during vegetative growth as the most important predictors while eliminating less influential variables [24]. The primary objective of LASSO is to minimize prediction errors while maintaining model interpretability. LASSO regression employs the L1 regularization method. In the current investigation, LASSO was implemented through the "glmnet" package within the R programming environment [4, 41–43]. The LASSO technique works by minimizing an objective function, which is defined as follows:

$$L_{lasso}\left(\hat{\beta}\right) = \sum_{i=1}^{n}\left(y_i - x'_i\hat{\beta}\right)^2 + \lambda\sum_{j=1}^{m}\left|\hat{\beta}_j\right| \quad (3)$$

Here:

- $\hat{\beta}$ denotes the regression coefficients associated with the explanatory variables in the LASSO model

- x refers to the input features (weather indices)

- y indicates the corresponding output or response variable (soybean yield)

- n indicates the total number of observations in the calibration dataset

- λ serves as the regularization or penalty hyperparameter

- m represents the number of predictor variables.

The optimal λ value for soybean yield prediction is typically determined through cross-validation, with studies showing that λ values between 0.01 and 0.1 often provide the best balance between model complexity and prediction accuracy.

#### *2.3.3.4 Principal component analysis-artificial neural networks models (PCA-ANN)*

In the PCA-ANN hybrid approach, the principal component (PC) scores were utilized as inputs for the neural network analysis [17]. To mitigate the problem of multicollinearity among weather-related variables, the PC scores were employed as predictor variables in the artificial neural network (ANN) to develop a robust soybean yield prediction model [44]. The PCA-ANN hybrid model has shown exceptional performance in soybean yield prediction due to its ability to combine the dimensionality reduction capabilities of PCA with the non-linear modeling strength of ANNs. Studies have demonstrated that PCA-ANN models can achieve $R^2$ values exceeding 0.85 for soybean yield prediction when properly calibrated. Principal Component Analysis (PCA) was employed to decompose the original data matrix X into two constituent matrices, T and P, represented as $X = TP^t$. Here, P is typically referred to as the loadings matrix, while T is an orthogonal matrix of scores. The superscript t indicates the matrix transpose. For soybean yield prediction, the first three to five principal components typically explain 80–90% of the total variance in weather data, making this approach both efficient and effective. The PCA-ANN model development process for soybean yield prediction involves several steps:

1. Standardization of weather variables

2. PCA transformation to extract principal components

3. Selection of components with eigenvalues >1

4. Neural network training using PC scores as inputs

5. Model validation and optimization.

This hybrid approach has proven particularly effective for soybean yield prediction in regions with complex weather patterns, where traditional linear models may fail to capture the intricate relationships between multiple weather variables and crop yield.

*2.3.4 Model performance evaluation*

The reliability of the prediction models was assessed using various statistical measures, including the Adjusted $R^2$ (coefficient of determination), *RMSE* (Root Mean Square Error), *nRMSE* (Normalized Root Mean Square Error), *MAE* (Mean Absolute Error), *PE* (Percentage Error), and Model Efficiency (EF). Higher values of Adjusted $R^2$ and *EF*, approaching 1, along with lower values of *RMSE* and *MAE*, closer to 0, and a narrower range of *PE*, indicate better model performance. Among the assessment indicators, Adjusted $R^2$ was predominantly utilized to compare model performance and was interpreted as follows: values above 0.90 were classified as excellent, between 0.75 and 0.90 as good, between 0.50 and 0.75 as fair, and below 0.50 as poor [45, 46]. Likewise, the normalized Root Mean Square Error (*nRMSE*) was applied to assess prediction quality: scores under 10% were labeled as excellent, between 10 and 20% as good, from 20 to 30% as fair, and above 30% as poor [5, 25, 29]. These performance metrics were computed using their conventional formulas:

$$\text{Adjusted R}^2 = 1 - \frac{(\text{n}-1)(1-\text{R}^2)}{(\text{n}-\text{p}-1)} \tag{4}$$

$$RMSE = \sqrt{\frac{\sum_{i=1}^{n}(y_i - \hat{y}_i)^2}{n}} \tag{5}$$

$$nRMSE = \frac{100}{M}\sqrt{\frac{\sum_{i=1}^{n}(y_i - \hat{y}_i)^2}{n}} \tag{6}$$

$$MAE = \frac{\sum_{i=1}^{n}|y_i - \hat{y}_i|}{n} \tag{7}$$

$$Percentage\ Error = \left(\frac{y_i - \hat{y}_i}{y_i}\right) * 100 \tag{8}$$

$$EF = \left[1 - \frac{\sum_{i=1}^{n}\left(y_i - \hat{y}_i\right)^2}{\sum_{i=1}^{n}\left(y_i - \bar{y}\right)^2}\right] \tag{9}$$

In this context, $y_i$ represents the actual (observed) value, while $\hat{y}_i$ denotes the estimated (predicted) value for each observation $i = 1, 2, 3, \ldots, n$. The notations $\bar{y}$ and $\hat{y}_i$ correspond to the mean values of the actual and forecasted data, respectively. Similarly, $\sigma_y$ and $\sigma_{\hat{y}}$ denote the standard deviations of the actual and predicted datasets. $M$ refers to the observed values, $n$ indicates the number of data entries or samples, and $p$ signifies the number of explanatory variables or input features used in the model.

## 3. Results and discussion

### 3.1 Model performance comparison

The comparative analysis of the three models revealed distinct performance characteristics across the eight districts of Uttarakhand. The PCA-ANN hybrid model consistently outperformed both the standalone ANN and LASSO models across all evaluation metrics ([3, 4, 23, 25, 26]; and [5]). The superior performance of PCA-ANN can be attributed to several factors:

1. *Dimensionality reduction*: PCA effectively reduced the dimensionality of the weather indices while preserving the most significant variance in the data, thereby eliminating redundant information and noise.

2. *Multicollinearity mitigation*: By transforming the original correlated weather variables into orthogonal principal components, the model avoided issues related to multicollinearity that could affect prediction accuracy.

3. *Enhanced generalization*: The combination of PCA's linear transformation with ANN's non-linear modeling capabilities provided modeling capabilities provided a robust framework for capturing complex relationships between weather patterns and crop yields.

4. *Reduced overfitting*: The dimensionality reduction achieved through PCA helped prevent overfitting, particularly important given the relatively small sample size typical in agricultural datasets.

### 3.2 Regional variations in model performance

The performance of the models varied across different districts, reflecting the diverse agro-climatic conditions within Uttarakhand. Districts with more stable weather patterns and consistent agricultural practices showed better model performance, while regions with high climatic variability presented greater challenges for yield prediction.

### 3.3 Weather index significance

The analysis of weighted versus unweighted weather indices revealed that correlation-based weighting generally improved model performance. This finding suggests that considering the historical relationship between weather variables and crop yields enhances the predictive capability of the models.

#### *3.3.1 Practical implications*

The developed models have significant practical implications for:

1. *Farm management*: Farmers can use yield predictions to make informed decisions about resource allocation, including fertilizer application, irrigation scheduling, and harvest planning.

2. *Policy development*: Government agencies can utilize these predictions for policy formulation related to crop insurance, minimum support prices, and food security measures.

3. *Market planning*: Agribusinesses can better plan their procurement and marketing strategies based on anticipated yield levels.

4. *Risk management*: Early warning systems can be developed to alert stakeholders about potential yield shortfalls or surpluses.

### **3.4 Limitations and future research directions**

#### *3.4.1 Data limitations*

The study faced certain limitations related to data availability and quality. The 20 years dataset, while substantial, could benefit from longer time series to capture more climate variability and extreme weather events. Additionally, the spatial resolution of weather data could be improved to better represent local microclimatic conditions.

#### *3.4.2 Model enhancements*

Future research could explore:

1. *Deep learning approaches*: Advanced neural network architectures such as Long Short-Term Memory (LSTM) networks or conventional Neural Networks (CNNs) could potentially capture temporal dependencies and spatial patterns more effectively.

2. *Ensemble methods*: Combining multiple models through ensemble techniques could improve prediction accuracy and robustness.

3. *Real-time integration*: Developing systems that can incorporate real-time weather data and satellite imagery for dynamic yield forecasting.

4. *Multi-crop models*: Extending the approach to other crops and developing integrated cropping system models.

#### *3.4.3 Technological integration*

The integration of emerging technologies such as Internet of Things (IoT) sensors, drone-based monitoring, and satellite imagery could provide additional data sources and improve model accuracy. Remote sensing data could particularly enhance the spatial resolution and temporal frequency of crop monitoring.

### 3.5 Conclusions

This comprehensive study successfully developed and evaluated three multivariate models for soybean yield prediction in Uttarakhand, with the PCA-ANN hybrid approach emerging as the most effective solution. The research demonstrated that:

1. *Hybrid models outperform individual techniques*: The PCA-ANN model's superior performance highlights the benefits of combining dimensionality reduction with non-linear modeling capabilities.

2. *Weather indices are crucial predictors*: The incorporation of phenological weather indices significantly improved model performance compared to using raw weather data.

3. *Regional adaptation is necessary*: Model performance varied across districts, emphasizing the need for location-specific calibration.

4. *Weighted indices enhance accuracy*: Correlation-based weighting of weather indices improved predictive capability.

The findings of this research contribute significantly to the field of agricultural forecasting and provide a robust framework for soybean yield prediction in Uttarakhand. The developed models can serve as valuable tools for farmers, policymakers, and agricultural stakeholders in making informed decisions related to crop management resources allocation, and food security planning. The methodology developed in this study can be adapted for other crops and regions, potentially contributing to broader applications in precision agriculture and sustainable farming practices. As climate change continues to pose challenges to agricultural productivity, such predictive models become increasingly important for maintaining food security and supporting agricultural sustainability.

## Acknowledgements

The authors acknowledge the NASA Langley Research Center POWER Project funded through the NASA Earth Science Directorate Applied Science Program.

## Author contribution

Yunish: Investigation, Formal analysis, Writing-original draft; Vinod Kumar: Supervision, Writing-review; Parul Setiya: Conceptualization, Methodology, Writing; Anurag Satpathi: Writing-review, Editing. All authors have read and approved the manuscript.

## Conflict of interest

The authors declare that they have no conflict of interest.

## Availability of data and materials

The datasets used in this study are available upon request.

## Code availability

The software used MS Excel and R Studio, and the scripts used in R Studio can be shared on request.

## Author details

Yunish Khan[1*], Vinod Kumar[1], Parul Setiya[2] and Anurag Satpathi[3]

1 College of Basic Sciences and Humanities, G.B. Pant University of Agriculture and Technology, Pantnagar, Uttarakhand, India

2 College of Agriculture, G.B. Pant University of Agriculture and Technology, Pantnagar, Uttarakhand, India

3 Division of Agrometeorology, Sher-e-Kashmir University of Agricultural Sciences and Technology, Kashmir, India

*Address all correspondence to: khanyunish263145@gmail.com

## References

[1] Mann ML, Warner JM, Malik AS. Predicting high-magnitude, low-frequency crop losses using machine learning: An application to cereal crops in Ethiopia. Climatic Change. 2019;**154**:211-227

[2] Cao J, Zhang Z, Luo Y, Zhang L, Zhang J, Li Z, et al. Wheat yield predictions at a county and field scale with deep learning, machine learning, and google earth engine. European Journal of Agronomy. 2021;**123**:126204

[3] Setiya P, Satpathi A, Nain AS, Das B. Comparison of weather-based wheat yield forecasting models for different districts of Uttarakhand using statistical and machine learning techniques. Journal of Agrometeorology. 2022;**24**(3):255-261

[4] Khan Y, Kumar V, Satpathi A, Setiya P, Surbhi K, Nain AK, et al. Comparative evaluation of hybrid and individual models for predicting soybean yellow mosaic virus incidence. Scientific Reports. 2025;**15**:15790. DOI: 10.1038/s41598-025-99427-5

[5] Satpathi A, Setiya P, Das B, Nain AS, Jha PK, Singh S, et al. Comparative analysis of statistical and machine learning techniques for rice yield forecasting for Chhattisgarh, India. Sustainability. 2023;**15**(3):2786

[6] Wu Q, Zhang K, Meng J. Identification of soybean leaf diseases via deep learning. Journal of the Institution of Engineers: Series A. 2019;**100**(4):659-666. DOI: 10.1007/s40030-019-00390

[7] Soystats. A reference guide to important soybean facts & figures. 2022. Available from: http://soystats.com/wp-content/uploads/Soy-Stats-2022-Web.pdf [Accessed: 20 April 2023]

[8] Bhartiya A, Chandra N, Pal RS, Aditya JP, Bajeli J. Comparative yield gaps, economic impact and constraint analysis in front line demonstrations of soybean and black soybean in Uttarakhand hills. Indian Journal Of Agricultural Research. 2017;**51**(5):483-487

[9] Banerjee P, Mukherjee B, Venugopalan VK, Nath R, Sarath Chandran MA, Dessoky ES, et al. Thermal Response of Spring–Summer-Grown Black Gram (*Vigna mungo* L. Hepper) in Indian Subtropics. MDPI; 2021. DOI: 10.3390/atmos12111489

[10] Garde YA, Dhekale BS, Singh S. Different approaches on pre harvest forecasting of wheat yield. Journal of Applied and Natural Science. 2015;7(2):839-843

[11] Ihsan MZ, El-Nakhlawy FS, Ismail SM, Fahad S, Daur I. Wheat phenological development and growth studies as affected by drought and late season high temperature stress under arid environment. Frontiers. 2016. DOI: 10.3389/fpls.2016.00795

[12] Ji Z, Yaozhong P, Xiufang Z, Jinyun W, Qiannan L. Prediction of crop yield using Phenological information extracted from remote sensing vegetation index. Sensors. 2021;**21**(4):1406. DOI: 10.3390/s21041406

[13] Ransing DM, Verma AK, Meshram MR, Paikra PR. Nodulation, yield and thermal requirement of mungbean (*Vigna radiata* L.) genotypes as influenced by date of sowing. International Journal of Agricultural Sciences. 2014;**10**(2):638-641

[14] Seo B, Lee J, Lee KD, Hong S, Kang S. Improving remotely-sensed

crop monitoring by NDVI-based crop phenology estimators for corn and soybeans in Iowa and Illinois, USA. Field Crops Research. 2019;**238**:113-128

[15] Ahmad T, Sahoo PM, Singh M, Biswas A. Crop cutting experiment techniques for determination of yield rates of field crops. 2021. 10.13140/RG.2.2.16327.62883

[16] Basso B, Cammarano D, Carfagna E. Review of crop yield forecasting methods and early warning systems. In: Proceedings of the First Meeting of the Scientific Advisory Committee of the Global Strategy to Improve Agricultural and Rural Statistics, FAO Headquarters, Rome, Italy. Vol. 18. 2013. p. 19

[17] Aravind K, Vashisth A, Krishanan P, Das B. Wheat yield prediction based on weather parameters using multiple linear, neural network and penalised regression models. Journal of Agrometeorology. 2022;**24**(1):18-25. DOI: 10.54386/jam.v24i1.1002

[18] Sridhara S, Manoj KN, Gopakkali P, Kashyap GR, Das B, Singh KK, et al. Evaluation of machine learning approaches for prediction of pigeon pea yield based on weather parameters in India. International Journal of Biometeorology. 2023;**67**(1):165-180

[19] Das B, Murgaonkar D, Navyashree S, Kumar P. Novel combination artificial neural network models could not outperform individual models for weather-based cashew yield prediction. International Journal of Biometeorology. 2022;**66**(8):1627-1638

[20] Sridhara S, Ramesh N, Gopakkali P, Das B, Venkatappa SD, Sanjivaiah SH, et al. Weather-based neural network, stepwise linear and sparse regression approach for rabi sorghum yield forecasting of Karnataka, India. Agronomy. 2020;**10**(11):1645

[21] Das B, Nair B, Arunachalam V, Reddy KV, Venkatesh P, Chakraborty D, et al. Comparative evaluation of linear and nonlinear weather-based models for coconut yield prediction in the west coast of India. International Journal of Biometeorology. 2020;**64**:1111-1123

[22] Heddam S, Vishwakarma DK, Abed SA, Sharma P, Ansari NA, Alataway A, et al. Hybrid river stage forecasting based on machine learning with empirical mode decomposition. Applied Water Science. 2024;**46**(14)

[23] Khan Y, Kumar V, Setiya P, Satpathi A. Comparison of phenological weather indices based statistical, machine learning and hybrid models for soybean yield forecasting in Uttarakhand. Journal of Agrometeorology. 2023a;**25**(3):425-431. DOI: 10.54386/jam.v25i3.2232

[24] Singh M, Nain AS. Statistical and machine learning approaches to study weather-disease-mustard yield relationship under varying environmental conditions. Research Square (United Kingdom). 2023:1-26. DOI: 10.21203/rs.3.rs-2459209/v1

[25] Khan Y, Kumar V, Setiya P, Satpathi A. Forecasting soybean yield: A comparative study of neural networks, principal component analysis and penalized regression models using weather variables. Theoretical and Applied Climatology. 2023b. DOI: 10.1007/s00704-023-04767-9

[26] Khan Y, Kumar V. Comparison of Phenological weather indices based statistical and machine learning models for soybean yield forecasting in Pantnagar, Uttarakhand. In: Proceedings of International Conference

on Mathematical Models, Statistics and Applications (ICMMSA-2023). 2023. pp. 1-21. DOI: 10.61463/ijset.icmmsa-2023.101

[27] Li X, Sha J, Wang ZL. Chlorophyll-a prediction of lakes with different water quality patterns in China based on hybrid neural networks. Water. 2017;**9**:524

[28] Rajaee T, Jafari H. Utilization of WGEP and WDT models by wavelet denoising to predict water quality parameters in rivers. Journal of Hydrologic Engineering. 2018;**23**

[29] Garde YA, Shukla AK, Singh S. Pre-harvest forecasting of rice yield using weather indices in Pantnagar (Uttarakhand). International Journal of Agricultural and Statistical Sciences. 2012;**08**(1):233-241

[30] Brejda JJ, Moorman TB, Karlen DL, Dao TH. Identification of regional soil quality factors and indicators, I. Central and southern high plain. Soil Science Society of America Journal. 2000;**64**:2115-2124

[31] Dalavi P et al. Modeling runoff in Bhima river catchment, India: A comparison of artificial neural networks and empirical models. Water Practice Technology (United Kingdom). 2024;**19**:2595-2612

[32] Elbeltagi A et al. Modelling daily reference evapotranspiration based on stacking hybridization of ANN with meta-heuristic algorithms under diverse agro-climatic conditions. Stochastic Environmental Research and Risk Assessment. 2022. DOI: 10.1007/s00477-022-02196-0

[33] Saroughi M et al. A novel hybrid algorithms for groundwater level prediction. Iranian Journal of Science & Technology Transactions of Civil Engineering. 2023;**47**:3147-3164. DOI: 10.1007/s40996-023-01068-z

[34] Dahikar SS, Rode SV. Agricultural crop yield prediction using artificial neural network approach. International Journal of Innovative Research in. Electrical, Electronics, Instrumentation and Control Engineering. 2014;**2**:683-686

[35] Kaul M, Hill RL, Walthall C. Artificial neural networks for corn and soybean yield prediction. Agricultural Systems. 2005;**85**(1):1-18

[36] Gupta S et al. Sensitivity of daily reference evapotranspiration to weather variables in tropical savanna: A modeling framework based on neural network. Applied Water Science. 2024;**14**:138

[37] Kushwaha NL et al. Stacked hybridization to enhance the performance of artificial neural networks (ANN) for prediction of water quality index in the Bagh river basin, India. Heliyon. 2024;**10**(10):e31085

[38] Shukla R et al. Modeling of stage-discharge using back propagation ANN-, ANFIS-, and WANN-based computing techniques. Theoretical and Applied Climatology. 2021. DOI: 10.1007/s00704-021-03863-y

[39] Kuhn M. Building predictive models in R using the caret package. Journal of Statistical Software (United States: Foundation for Open Access Statistics). 2008;**28**(5):1-26

[40] Kumar S. Comparison of linear regression and artificial neural network technique for prediction of a soybean biodiesel yield. In: Energy Sources, Part A: Recovery, Utilization, and Environmental Effects. 2019. DOI: 10.1080/15567036.2019.1604858

[41] Hastie T, Qian J, Tay K. An Introduction to glmnet. United States: Stanford University; 2021. Available from: https:/glmnet.stanford.edu/articles/glmnet.html

[42] Satpathi A et al. Estimation of crop evapotranspiration using statistical and machine learning techniques with limited meteorological data: A case study in Udham Singh Nagar, India. Theoretical and Applied Climatology. 2024. DOI: 10.1007/s00704-024-04953-3

[43] Satpathi A et al. Evaluating statistical and machine learning techniques for sugarcane yield forecasting in the Tarai region of North India. Computers and Electronics in Agriculture. 2025;**229**:109667

[44] Verma U, Piepho HP, Goyal A, Ogutu JO, Kalubarme MH. Role of climatic variables and crop condition term for mustard yield prediction in Haryana. International Journal of Agricultural and Statistical Sciences. 2016;**12**:45-51

[45] Joshi B et al. A comparative survey between cascade correlation neural network (CCNN) and feedforward neural network (FFNN) machine learning models for forecasting suspended sediment concentration. Scientific Reports. 2024;**14**:10638

[46] Vishwakarma DK et al. Modeling of soil moisture movement and wetting behavior under point-source trickle irrigation. Scientific Reports. 2023;**13**:14981

Section 4

# Socio-economic and Adoption Perspectives

**Chapter 7**

# Farmers' Responses to Soybean (*Glycine max (L.) Merrill*) Farming and the Influencing Socio-economic Factors in Jember District, Indonesia

*Syamsul Hadi, Insan Wijaya and Anisa Nurina Aulia*


## Abstract

Rapid population growth, shrinking agricultural land, and farmers' weak response to soybean cultivation have increased dependence on imported soybeans. Although soybean productivity in Jember performed better than the provincial average, particularly before 2020, a significant decline occurred in 2020, likely due to external factors such as climate conditions, land-use changes, or input constraints. Despite gradual improvement afterward, productivity has not returned to its 2017 peak. This study aims to (1) identify the characteristics of soybean farmers, (2) analyze the profitability and feasibility of soybean farming, (3) assess farmers' responses to soybean cultivation, and (4) determine the socio-economic factors causing their weak response to cultivating soybeans on controlled land. The research applies a descriptive quantitative and qualitative approach using a survey and summative technique. Samples include farmers who plant and those who do not plant soybeans. Primary and secondary data were analyzed using descriptive analysis, Likert scale scoring, profit analysis, and the R/C ratio. Results indicate that the average farmer age in Jember Regency is 49.26 years (productive age) with 9.42 years of formal education (moderate level). The average household size is 3.26 persons. The average soybean land area is 0.55 hectares, while total controlled paddy land averages 0.36 hectares. Soybean farming generates an average profit of Rp 4,487,125.76 per hectare with an R/C ratio of 1.83, indicating feasibility. However, farmers' overall response remains weak due to lower profitability compared to other commodities and the absence of government price protection.


**Keywords:** farmer response, soya commodity, profit, business feasibility, socio-economic factors

## 1. Introduction

The Ministry of Agriculture of the Republic of Indonesia has implemented an important strategy to reposition the agricultural sector as the driving force of national development, especially in rural areas, towards achieving food self-sufficiency including rice, corn and soybean commodities. The soybean commodity is one of the main commodities that contain vegetable protein sources and is used as a food base in agro-industrial sub-sector activities such as tofu, tempeh, soy sauce, tauco, oncom and soy milk. The current problem of soybeans in Indonesia is the decreasing domestic production, so it still relies on imported soybeans. On the one hand, domestic soybean production in 2023 increased to 301,520 tonnes, but the demand reached 2,800,000 tonnes, which has implications for the high dependence on imported soybeans [1].

Hulu's [2] view states that along with the gap between the amount of soybean demand and its production, the import policy must be able to suppress the increase in domestic soybean prices so that the microeconomic growth rate will grow well. This condition has been seriously highlighted by Santosa [3] who conducted a comparative study of the production costs of soybean farming with imports. If the price of imported soybean is IDR 1500 per kilogramme, while the price of local soybean at the farm level is IDR 2500 per kilogramme, the price of local soybean is under pressure. Therefore, these conditions will result in a tendency for farmers to leave soybean cultivation and switch to other commodities slowly.

East Java Province is the province with the largest soybean production in Indonesia in 2022 (27.96%). However, in the same year, the need for soybeans in East Java Province reached 267,000 tons, while local soybean production only reached 70,000 tons (26.22%), and the rest was met by imported soybeans, even in 2020, the need for soybeans was only met by 12.78% of local soybeans. This condition is caused by limited production capacity and losing out to imported soybeans so farmers tend to plant rice and corn, especially since they have not received incentives from the government. When compared to national conditions, it turns out that Indonesia is the 9th largest soybean importer with a total import of 2.65 million tons (USDA 2022 in Satriani et al. [4]). According to Risandi (2022) in Satriani et al. [4] that this condition is caused by factors such as relatively low local soybean productivity, the technology adopted is still traditional, business actors' preferences have different needs, the quality of local soybeans is still low, the price of local soybeans is more expensive than imported soybeans. The average price of imported soybeans in Indonesia is around IDR 14,200 per kg (as of September 2022).

East Java Province is the national soya food barn and a number of its regions are Gresik, Banyuwangi, Madiun, Nganjuk, Jember, Bojonegoro, Kediri, Blitar, Trenggalek and Lamongan. Among these areas, Jember Regency experienced a significant decline in production despite having the fifth highest productivity level in East Java. Along with a soybean planting area of around 7500 hectares, in 2021 Balitbangtan through Balitkabi chose Jember Regency as one of the bases for the development of high productivity soybean cultivation technology innovations. One of the leading technologies disseminated is the planting of superior varieties (Balitkabi 2021 in [5]). Comparing the development trend of total soybean production in East Java Province and Jember District is quite interesting because it has a relatively similar trend, but the average productivity of Jember District exceeds the average in East Java (**Figures 1** and 2).

The results of research by Hazmi et al. [6] showed that the growth rate of soybean-based agribusiness in the districts of Jember, Banyuwangi and Pasuruan from 2011 to 2016 decreased on average below 10% at least in terms of the number of business

*DOI: http://dx.doi.org/10.5772/intechopen.1011099*

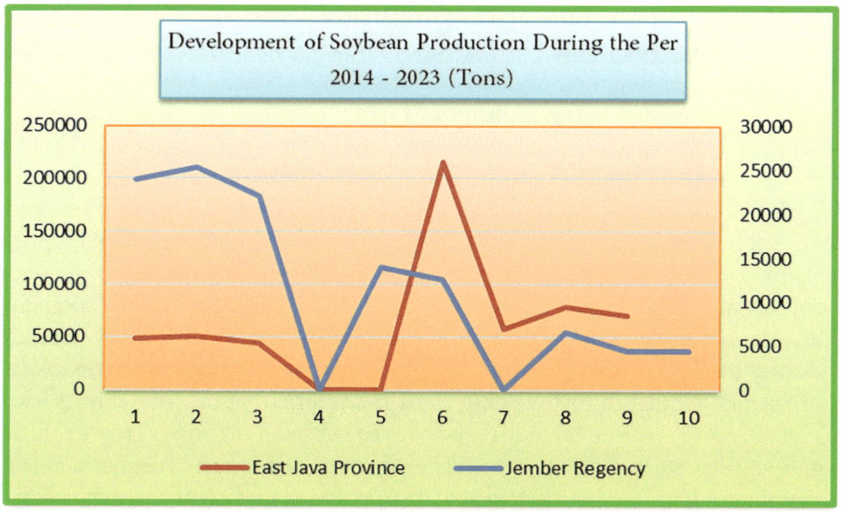


**Figure 1.**
*Fluctuations in the development of soybean production amounts in East Java province and Jember regency (Source: Central Bureau of Statistics (BPS) of East Java Province, 2024).*

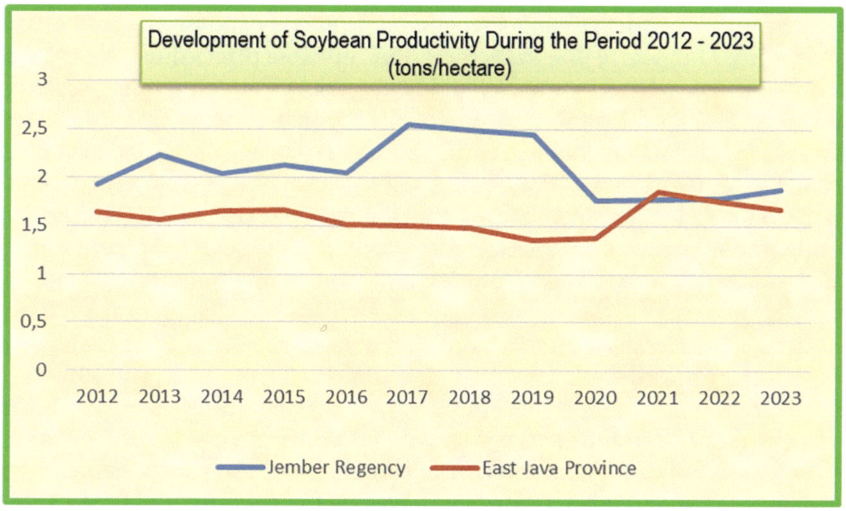


**Figure 2.**
*Development of soybean productivity in East Java Province and Jember Regency (Source: Central Bureau of Statistics (BPS) of East Java Province, 2024).*

units. This is due to the decreasing availability of soybean raw materials and the high price of soybean raw materials (an average increase of 25.99% per kg). This phenomenon is caused by the failure of the implementation of the PUAP Programme, which is indicated by the performance of programme actors from the soybean farming process to the processing of products averaging 44.76%. This shows that the role of Gapoktan in running PUAP is less than optimal in encouraging Poktan members to plant soybeans. In line with the less than optimal role of Gapoktan, this will also result in a weak response from farmers to be interested in planting soybean commodities, which has a further impact on the low participation of farmers to plant them.

The development of the number of soybean production and productivity conditions as shown in **Figures 1** and **2** is disproportionate when compared to the area of paddy fields in Jember Regency which is recorded at 84,328.30 hectares. Similarly, farmer institutions recorded the number of farmer groups has reached 1713 in the working area of 248 Gapoktan, but there are only 110 farmer groups (6.39%) that plant soybean commodities every dry season [7]. Since 2018, farmer groups whose members planted soybean commodities were spread across 22 sub-districts and reduced to 8 sub-districts in 2024. This phenomenon indicates the weak response and participation of farmers to persist in planting soybean commodities [8].

The constellation of unrest is in line with the results of research by Zakaria [9] and Widowati and Alfina [10], which revealed that the level of farmers' participation independently to meet the needs of soybeans is still low (4.21%). This means that the role of farmer groups in influencing their members to plant soybeans is less than optimal, resulting in weaker farmer responses to soybean farming. However, this condition is significantly different from the phenomenon of the results of research by Noermansyah and Dini [11], which revealed that the response of farmers participating in the SL-PTT Soybean Programme in Ciamis Regency is generally in the high category. This means that the potential response and participation of farmers to grow soybeans in any region including in Jember District, East Java Province is very potential. Therefore, it is necessary to mobilise farmers' participation in the form of farmer groups, thus strengthening farmers' participation to plant soybean.

The condition of the decreasing soybean planting area and production in Jember Regency (**Figure 1**) is partly due to the factor of unstable prices due to the flood of imported soybeans with better quality and lower prices. In fact, Jember District, which is part of East Java Province, is one of the soybean food barns along with Banyuwangi, Pasuruan, Lamongan, Sampang and Ponorogo Districts. East Java Province is still a mainstay with a contribution to national production of 42%, with a harvest area reaching 41% in 2024. The facts in the field show that the area planted and the amount of soybean production in several districts such as Blitar, Ponorogo and Lamongan are experiencing a significant decline, while Banyuwangi and Sampang and Bojonegoro are experiencing an increase [12]. Farmers are shifting their efforts to other commodities that are considered more economically prospective, such as citrus and dragon fruit.

Looking back at the previous decades, in an effort to maintain its contribution as the highest food buffer in Indonesia, East Java Province launched the development of soybean and rice areas in Jember District starting in 2013. In 2012 alone, the total soybean planted area in Jember District reached 15,233 hectares or 6.84% of the total planted area in East Java, which reached 222,738 hectares. The soybean planting area in the Jember Regency is fourth after Banyuwangi, Lamongan and Sampang Regencies. However, Jember Regency, over the last 5 years has experienced sharp fluctuations in planted area, although its productivity tends to increase quite well [13].

There are several problems in the development of soybean commodities, including the current climate/weather, land development specifications, sources and availability of superior seeds, biotic stresses, how to control them and the response of farmers to grow soybean tends to decline. The absence of government protection and guarantee of favourable prices for soybean commodities has implications for the lack of response of farmers to cultivate soybean crops. Whereas soybean production in East Java in 2023 is still in the range of 337 thousand tonnes, while the soybean consumption needs of East Java people are around 420 thousand tonnes per year. Because soybean production is still minus compared to consumption needs, East Java still has to import from outside [12]. Based on this background, the objectives of this study are (1) to identify the characteristics of soybean farmers, (2) to calculate the profit level and feasibility of soybean farming, (3) to measure farmers' response to soybean farming and (4) to determine the socio-economic factors that cause farmers' response to be weaker to cultivate soybean commodities on the land area controlled.

## 2. Research methods

### 2.1 Types and methods

This type of research is descriptive research, which is research that aims to describe events (phenomena) systematically, factually and accurately about the facts, characteristics and relationships between phenomena that occur in the present. Meanwhile, the method used in this research is the *survey* method where this method determines the sample taken from part of the existing population [14]. Meanwhile, to strengthen the first method, other methods are also used in this research, including documentation by studying books and or literature, research results, written records and so on that are relevant to the purpose of this research [15].

### 2.2 Research location and time

This research has been conducted in several sub-districts that are the centres of soybean production in Jember Regency. The research locations were determined by *purposive sampling* based on the consideration that these sub-districts are the first to fifth largest soybean production centres in Jember Regency in 2023, including the sub-districts of Bangsalsari, Balung, Umbulsari, Puger, Rambipuji. In addition, this research has also been conducted in sub-districts that have been the centre of soybean production before the research year as a control which includes the sub-districts of Jelbuk, Kalisat, Ambulu, Mayang and Tempurejo [7].

### 2.3 Sampling and data collection techniques

The determination of the sample of respondents was taken by simple random sampling of part of the existing population of both the population of farmers who are still planting soybean commodities and no longer cultivating them. The data collection techniques were carried out by interviewing respondents through *depth interview* and Focuss group discussion (FGD) techniques or other techniques that are easier and more practical to use and complement each other [16].

### 2.4 Data analysis

In order to identify the characteristics of soybean farmers in this research location, descriptive analysis was used, and the results were presented in the form of figures. As for calculating the level of profit and economic feasibility of soybean farming, each is analysed with the formulation of the profit formula [17] and the formulation of the R/C ratio formula [18] along with the decision-making criteria. If the R/C ratio value is greater than (1 + bank interest rate), then the soybean farm is feasible, and if it is smaller or equal to (1 + bank interest rate), then there is no need to continue. Furthermore, to measure the strengths and weaknesses of farmers' responses to soybean farming, *a scoring analysis* with a *Likert* scale approach was used. The consideration is based on the development of a scale to measure farmers' attitudes towards a particular aspect known as the Likert scale [14]. Based on the measurement of response indicators, which include perception, motivation, attitude, skills and participation of farmers towards soybean commodity farming as determined as follows: If the total score is between 84 and 100 (very strong farmer response), between 68 and 83 (strong farmer response), between 52 and 67 (less strong farmer response), between 36 and 51 (weak farmer response) and between 20 and 35 (very weak farmer response). Furthermore, in order to inventory the socio-economic factors that cause the weak response of farmers to plant soybean commodities, descriptive analysis was conducted using the frequency table method in units (%).

## 2.5 Characteristics of soybean farmers

**Figure 3** shows that the average age of farmers in Jember District is 49.26 years (classified as productive age), with a range of 22–71 years. This indicates that the farmers are in the productive age (15–64 years) according to the *International Labour Organisation* (ILO) and Sugiana et al. [19]. The age of a person in this group is physically and mentally able to work and do business optimally. It should be noted that according to the 2023 Agricultural Census data from BPS, the average age of farmers in Indonesia is currently above 55 years old. This means that farmers in Jember Regency actually have higher productivity compared to the national average because they have stronger physical conditions and more energy.

**Figure 3** also illustrates that more than 90% of farmers have adequate physical strength and mental stability, so they tend to be able to manage their businesses well. At this mature age, farmers are generally able to consider exactly what commodities need to be cultivated on the land they control in order to obtain maximum farm profits and on the other hand to support major national food security programmes such as soybean food. Among farmers in Jember District, those classified as millennial farmers (aged 19–39 years) are 20% and lower than the national average of 6,183,009 people, or 21.93% of total farmers.

**Figure 4** shows that the average formal education level of farmers is medium (9.42 years) with a range between 6 and 16 years. When compared to the average level of education of farmers in Indonesia (elementary and junior high school graduates),

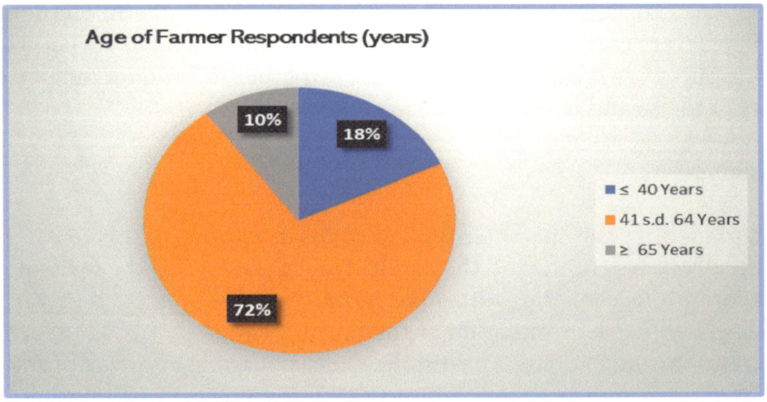


**Figure 3.**
*Distribution of farmers by age in Jember Regency in 2024.*

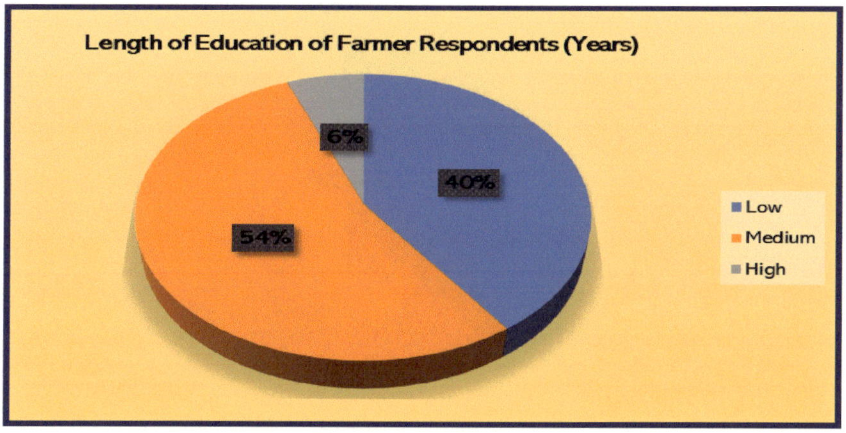


**Figure 4.**
*Distribution of farmers by years of formal education in Jember regency in 2024.*

the average level of education of farmers in Jember Regency is currently slightly higher, namely junior high school to senior high school graduates. Indeed, with this level of education, farmers in the Jember Regency should be more advanced in managing the land they control, at least to encourage a higher level of land productivity compared to the average in other regions. **Figure 4** also shows that the majority (54%) of farmers in Kabupaten Jember have a medium level of education, 40% have a low level of education and only 6% have a high level of education. The condition of this level of education will have implications for the way attitudes and decision-making are determined, especially in determining the commodity to be cultivated in accordance with the appropriate planting season and on consideration of other factors such as past season experience, estimated input and output prices.

The condition of the farmer's education level as described above also reflects a fairly broad mastery of science that has implications for rationality in thinking, decision-making and performance. However, the knowledge does not have to be solely obtained from formal education but can be obtained from non-formal education such as internships, training, seminars and workshops. This is in accordance with Soekartawi's [20] opinion that the level of formal education is one of the factors that influence a person to think in a better and rational direction.

Furthermore, **Figure 5** shows that the average number of farmer household members in Kabupaten Jember is 3.26 people (classified as a medium family), with a range between 1 and 6 people per household. When compared to the average number of dependents in farmer households nationally (around four people), the condition in Kabupaten Jember is still lower. The size of the household will determine the amount of labour that will be allocated to farming activities (own and others' farms) and non-farming activities. However, this labour factor, according to the results of research by Soetriono et al. [21], has no significant effect on the amount of soybean production in East Java Province.

Furthermore, regarding the number of dependents of farmer households, it can be explained in more detail that most (52.83%) of the dependents of farmer households in Kabupaten Jember are between three and four people, 28.30% are between one and two people and the rest are as many as five people. When compared to the national condition, it can be described that 58% of farmer households have one to two dependents, 40% of farmer households have three to four dependents and 2% have more than five dependents (**Figure 5**).

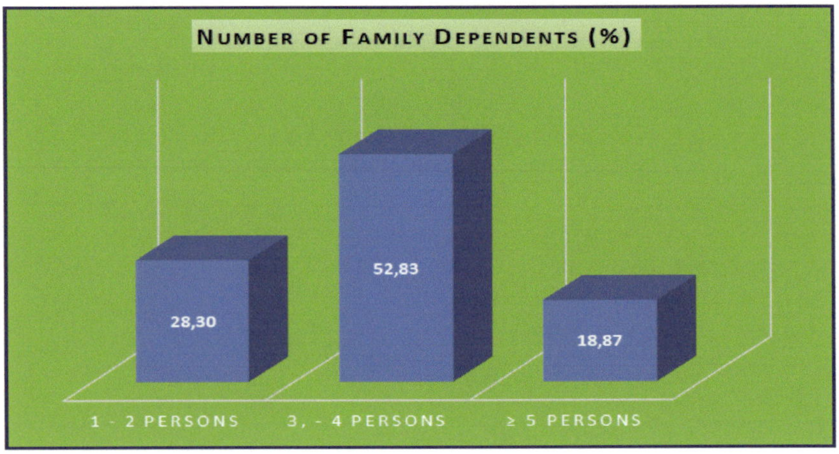


**Figure 5.**
*Distribution of farmers by number of household members in Jember Regency in 2024.*

In relation to the number of household dependents, Mantra [22] argues that families with between four and five family members are classified as moderate families. In general, families with more family members tend to affect the motivation to work with the aim of fulfilling and prospering their family members, including the provision of labour to support the business being run. In addition, the number of household dependents will affect the availability of labour and opportunities to work in the agricultural sector. It should be noted that as of August 2024, the agriculture, forestry and fisheries sectors absorbed 28.18% of all Indonesian workers. This condition will also affect farmers' considerations or decisions to determine which commodities are rational to cultivate on the land they control.

**Figure 6** also illustrates that the average land area controlled by farmers and planted with soybean commodities is 0.55 hectares with a range of 0.1 to 3 hectares. Most (70%) farmers planted soybeans on land units ≤0.5 hectares, some (20%) between 0.6 and 1 hectare and 10% on more than 1 hectare. This condition can be explained by the fact that the area of paddy fields planted with soybean

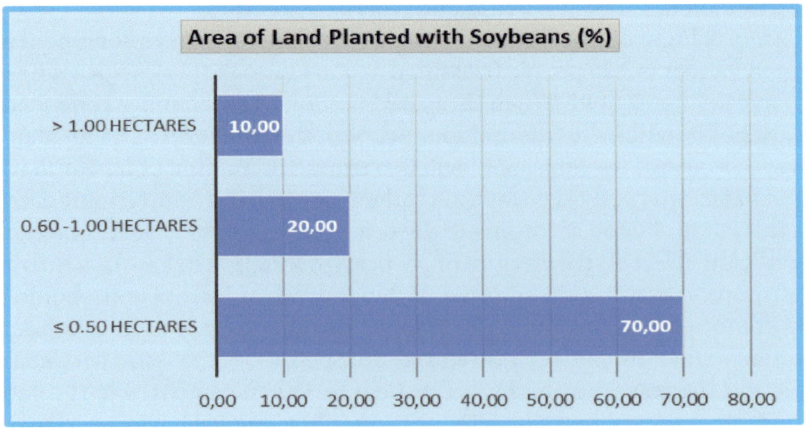


**Figure 6.**
*Distribution of farmers by land area planted with commodities soybean in Jember Regency in 2024.*

commodities on the average total area of paddy fields controlled by farmers (= 0.96 hectares) is only 29.33%. This is because farmers tend to reduce the risk of decreasing soybean farming profits due to various factors such as the level of land productivity, the availability of plenty of irrigation water, the agro-climate that is not suitable for soybean cultivation, the absence of economic incentives from the government and the estimated selling price of soybean production. However, there are 40% of farmers who plant soybeans on the entire land area controlled due to agro-climatic factors suitable for planting rice and being a pioneer for other farmer members.

The narrowness of the overall average size of paddy fields controlled by farmers in Kabupaten Jember, which is 0.36 hectares, is in line with the high rate of population growth in the agricultural sector. Therefore, the average unit area owned by farmer households becomes narrower along with the average number of dependents or household members. This condition is in line with the national phenomenon, where data from BPS shows that in 2023, 68.10% of agricultural businesses in Indonesia were categorised as small-scale farmers or the land area controlled was ≤0.5 hectares [23]. Thus, according to Soekartawi [17], the average farmer in Java is classified as a smallholder with a cultivated land area of <0.5 ha. The implication is that along with the narrow land area controlled by farmers, it will further result in the opportunity to plant soybean commodities tends to decrease, especially if some other inhibiting factors are still a *constraint* in encouraging farmers to participate in planting soybean commodities during the season.

Meanwhile, in the context of the length of experience in soybean farming, the average experience of farmers in Jember Regency in soybean farming has lasted more than 25 years. The length of time of experience is relatively long for a person's existence in running his business. This business experience greatly determines the success of a person in running or managing his business because the longer a person's experience, the stronger the level of skill. In fact, no less than 23% of farmers have not cultivated their land by planting soybean commodities for an average of 5 years, with a range of 5–10 years.

This condition is in line with farmers in Banyuwangi Regency in the Southern part of East Java Province. As many as 50% of farmers in the area have not planted soybean commodities on rice fields, yards and moorlands. Farmers have shifted their land management to other commodities such as dragon fruit, citrus, tobacco and other horticultural commodities. Furthermore, the average scale of soybean farming both in respondents who still persist in planting soybeans is 0.49 ha, planting citrus fruits is 1.1 ha and the average planting of dragon fruit is 4.68 ha, consisting of paddy fields, moorlands and yards. Even as many as 28.95% of respondents of citrus and dragon fruit farmers are forced to rent land owned by others with the intention of expansion. In addition, not all land owned by respondents is planted with soya, given the uncertain climate, falling soya output prices and the trauma of the previous season [6].

The results of research by Hazmi et al. [6] also revealed that the average orange and dragon farmers in the Southern District of Banyuwangi have left soybean farming since 13 years ago. They began to look at other commodities (citrus fruits and dragons) after soybean commodities were no longer attractive to cultivate due to factors of fluctuations in output prices and farmers often lost money as well as economically prospective factors for citrus fruits and dragons by looking at the number of demands for both commodities increasing from time to time with very tempting prices.

## 2.6 Soybean farming analysis

### *2.6.1 Cost structure of soybean farming*

The results revealed that the average total fixed costs in soybean farming in the study area reached 45.56% of the overall farming costs. The high allocation of fixed costs is due to the fact that all land cultivated by soybean crops is calculated as rent, so the allocation for rental costs in soybean farming in the study area reaches 37.45%. Details of the structure of soybean farming costs per hectare in the study area are presented in **Table 1**.

**Table 1** shows that the allocation of variable costs in soybean farming in the study area reached 54.44%, of which most (37.44%) were labour costs. The use of labour is actually still below the recommendation (90–100 MWD) according to the Directorate of Cultivation of Beans and Tubers – Directorate General of Food Crops, Ministry of Agriculture of the Republic of Indonesia [24]. This condition is due to the fact that the cultivation techniques applied are still far from the recommendation, so the use of labour only reaches 70.13 MWD per hectare. The relatively low labour allocation in

| No. | Cost description | Number of respondents (people) | Unit | Volume | Rp/Unit | Value (Rp) | Percentage (%) |
|---|---|---|---|---|---|---|---|
| A | Variable costs: | | | | | | |
| 1 | Means of production: | | | | | | |
| | a. Seeds | 50 | Kg | 44.16 | 10,012.00 | 442,128.80 | 7.89 |
| | b. Urea fertiliser | 19 | Ku | 0.16 | 300,000.00 | 48,100.00 | 0.86 |
| | c. TSP | 11 | Ku | 0.10 | 216,363.64 | 21,982.55 | 0.39 |
| | d. NPK | 28 | Ku | 1.02 | 216,666.67 | 221,372.04 | 3.95 |
| | e. Medicines | 27 | Ls | 3.15 | 25,777.78 | 81,261.38 | 1.45 |
| 2 | Labour | 50 | MWD | 70.13 | 29,917.68 | 2,098,013.47 | 37.44 |
| | *Sub Total* | | | | | *2,912,858.22* | 54.44 |
| B | Fixed Costs: | | | | | | 0 |
| 1 | Tool depreciation cost | 50 | Ls | 1.00 | 237,601.83 | 237,601.83 | 4.44 |
| 2 | Land tax | 36 | Season | 1.00 | 127,833.89 | 127,833.89 | 2.39 |
| 3 | Land rent | 50 | Season | 1.00 | 2,003,588.24 | 2,003,588.24 | 37.45 |
| 4 | Irrigation fee | 19 | Season | 1.00 | 17,126.61 | 17,126.61 | 0.32 |
| 5 | Group dues | 21 | Ls | 1.00 | 51,242.86 | 51,242.86 | 0.96 |
| | *Sub total* | | | | | *2,437,393.43* | 45.56 |
| | Total | | | | | 5,350,251.66 | 100 |

*MWD, men's working day; Ls, and lump sum. Source: Primary Data Processed in 2024.*

**Table 1.**
*Structure of soybean farming costs per hectare in Jember regency in 2024.*

this case can be shown as only 38% of respondents used urea fertiliser, 22% of respondents used TSP fertiliser, 52% of respondents used drugs such as biosem, rempit and gundasil at a dose of between 1.14 litres/cc per hectare. The rest did not spray because they did not see any symptoms of pest and disease attack.

Furthermore, the consideration of farmers not to spray soybean farming plants in addition to limited capital is also in accordance with the recommendations in the 2013 Technical Guidelines for Soybean Production Management issued by the Directorate of Bean and Tuber Cultivation – Directorate General of Food Crops of the Ministry of Agriculture of the Republic of Indonesia that the application of effective insecticides is adjusted to the needs, namely according to the intensity of the attack or pest population based on observations or if it has reached the economic threshold, then spraying with insecticides is carried out. However, the total cost of soybean farming in 2024 is between IDR 4,500,000 and IDR 9,800,000 per hectare. The largest component of the production cost of soybean farming is for land rental expenditure of 40.34% (although owned land is counted as rent) of the total cost. Meanwhile, the use of soybean seeds on the farm is in accordance with the recommendation (40–50 kg/ha), which is 44.16 kg per hectare, but the average use of soybean seeds in East Java Province exceeds the recommendation [21].

In addition to the above reasons for the lack of labour use, some farmers (54%) applied No-Tillage (TOT) because the previous season was planted with rice. Therefore, land preparation work is only carried out to clean up rice waste/straw, make small manholes for irrigation water drainage and clean the rice field bunds. In addition to limited capital, another factor is the scarcity of labour in rural areas, making it difficult to find labour when activities in soybean farming are needed. Farmer respondents partly use labour within the family and are counted as non-cash expenses in the cost analysis. This condition is in line with the results of Widotono and Arifin's [25] research in East Java Province that the amount of labour in soybean farming in the research area is mostly sourced from within the family only, on the one hand, and on the other hand, aggravated by the scarcity of labour in the agricultural sector.

#### *2.6.2 Soybean farming profits*

The level of productivity of soybean farming in Jember District reached 1.38 tonnes per hectare, while the average productivity in East Java reached in 2023 a fixed figure of 14.56 quintals/hectare. However, it is expected that in 2024 it will increase to 1.7–2 tonnes per hectare. This means that the productivity of soybean farming in Jember Regency is still below the average in East Java Province and when viewed from the national target with a soybean farming productivity level of 2.1 tonnes/ha, the productivity achievements in Jember Regency still need to be re-evaluated. According to the national standards set by the Directorate of Cultivation of Various Beans and Tuber – Directorate of Food Crops - Ministry of Agriculture of the Republic of Indonesia in 2023, the level of productivity of soybean farming above 2 tonnes/ha is classified as high, between 1.5 and 2 tonnes/ha and less than 1.5 tonnes/ha is classified as low/low.

**Table 2** shows that the average price of soybean in the study area was IDR 7136 per kg with a range of IDR 6000 to IDR 9000 per kg and the average production was 1380.28 kg per hectare with a range of 1167–1520 kg. Furthermore, the average total cost of production is IDR 5,352,377 per hectare, with an average soybean farming revenue of IDR 9,839,502.41 per hectare, so that the average profit reaches IDR 4,487,125.76 per hectare. This average level of profit is relatively not high when compared to the cost sacrifices incurred by farmer respondents. This is because the

| No. | Description | Unit | Value (IDR) |
|---|---|---|---|
| 1 | Production | Kg | 1380.28 |
| 2 | Variable costs | IDR | 2,942,437.06 |
| 3 | Fixed costs | IDR | 2,437,393.43 |
| 4 | Production price | IDR/Kg | 7136.00 |
| 5 | Production value | IDR | 9,839,502.41 |
| 6 | Advantages | IDR | 4,487,125.76 |
| 7 | ROI (rate of return on capital) | % | 83.41 |

*Source: Primary data processed in 2024.*

**Table 2.**
*Soybean farming profits per hectare in Jember regency in 2024.*

level of production per hectare is still relatively low and has not been supported by adequate prices. The government has not set a *floor price*, so the selling price at the farm level is relatively low because the price of soybean commodities is determined by traders. The low level of profit is also supported by data from the Department of Food Crops, Horticulture and Plantations (DTPHP) of Jember Regency [23] which informs that the range of farm profit levels in Jember Regency ranges from IDR 4,000,000 to IDR 15,000,000 per hectare.

## 2.7 Economic feasibility analysis of soybean farming

In order to determine the extent to which soybean farming is economically feasible or not, a cost-efficiency analysis tool can be used by comparing the revenue with the costs sacrificed. **Table 3** illustrates that the average R/C ratio in the research area per cultivated land area and per hectare is 2.38 and 1.83, respectively. This means that every cost incurred, as much as IDR 1, will generate revenue of IDR 1.83 per hectare. This indicates that soybean farming is economically feasible to be implemented or continued. Although the average R/C ratio is only 1.83, as many as 46% of farmer respondents have reached the R/C ratio level above 2. Regarding the results of the analysis of the economic feasibility or cost efficiency of soybean farming with the R/C ratio analysis approach in the research area can be seen in **Table 3**.

The average value of R/C ratio in the research area is still higher when compared to the results of research by Feriady in 2013 in Bengkulu and Zakaria et al. in 2010 [26] in three provinces in Indonesia, each of which only reached 1.49. Furthermore, the R/C ratio value of soybean farming in the research area is relatively the same as the

| No. | Description | Per unit area | Per hectare |
|---|---|---|---|
| 1 | Production value (IDR) | 3,769,570.00 | 9,839,502.41 |
| 2 | Production cost (IDR) | 1,587,149.86 | 5,379,830.49 |
| 3 | R/C Ratio | 2.38 | 1.83 |

*Source: Primary data processed in 2024.*

**Table 3.**
*Level of R/C ratio of soybean farming in Jember regency in 2024.*

results of research by Sulastri et al. in 2011 in Sukorejo District, Ponorogo Regency and higher than the results of research by Meryanti [27] in Ciranjang District, Cianjur Regency with R/C ratio values of 1.82 and 1.35, respectively. However, when compared to the results of a study by the Aceh Food Security and Agricultural Extension Service in collaboration with the NAD Agricultural Technology Assessment Center in 2008 that analysed soybean farming of the anjasmoro variety in paddy fields which revealed that the R/C ratio of soybean farming reached 3.23, the R/C ratio of soybean farming in this research area is still far below.

The condition of economic feasibility is caused by several factors, among others: (1) the soybean seed assistance programme of superior varieties in the research area is uneven; (2) assistance by field officers is not optimal so that the process of implementing soybean cultivation is not intensive and far from the recommendations and (3) farmer respondents conduct soybean farming only driven by tradition or culture of planting in the dry season or Second Rainy Season regardless of the guarantee of output prices and availability of inputs that are uncertain and do not seem to have a strong motivation to be able to do business optimally due to government attention and protection felt very low.

Although the level of profit and the level of R/C ratio of soybean farming in the study area are not yet optimal, it is still economically feasible to continue considering the broad market prospects. This is because the development trend of the amount of production and consumption of soybean commodities in Indonesia in recent years has experienced a deficit of up to 1,301,000 tonnes in 2023 and the development of the amount of soybean production in East Java for the period 2015–2024 has decreased by 6.67%. In addition, the potential is supported by the national phenomenon that in the same period, there is a downward trend in soybean consumption much lower than the decline in production.

The implication of the soybean commodity deficit phenomenon is that without a strategy to increase soybean production, Indonesia will face an increasing deficit. This opportunity is also supported by the results of research by Feriady in Bengkulu that the investment in soybean farming under normal circumstances (as projected) has a positive NPV value of Rp 4,908,380.02, Net B/C = 2.78 and IRR = 73.82% which is far above the percentage of loan interest (24% per year). This implies that the project is very feasible to implement. The sensitivity analysis results show that the project is highly sensitive to a 10% drop in revenue and a one-year production delay. This situation can be seen from the negative NPV, net B/C < 1 and IRR < 24% or less than the percentage of loan interest (24% per annum).

### 2.8 Farmer response and the socio-economic factors that influence it

Farmers' response to soybean farming in Jember Regency has a tendency to weaken, so that the level of participation of farmers to plant soybean commodities also tends to be lower. This is supported by the results of the study which revealed the average total score of the four response indicators which include perception, motivation, knowledge and attitude is 63.72. This average score is classified as a less strong or weak response group that has a score between 52 and 67 based on predetermined decision criteria with a maximum total score of 110. The complete distribution of the average score of farmers' responses to soybean commodity farming is presented in **Table 4**.

**Table 4** illustrates that from the total average score of farmer responses (63.72), it turns out that the highest average score is in the aspect of farmer perception (29.02), which has a score interval between 9 and 45. This means that for the perception

| No. | Farmer response description | Average total score |
|---|---|---|
| 1 | Farmer perception | 29.02 |
| 2 | Farmer motivation | 10.46 |
| 3 | Farmer knowledge | 13.06 |
| 4 | Farmer attitudes | 11.18 |
| Total Average Score | | 63.72 |

*Source: Primary Data Processed in 2024.*

**Table 4.**
*Respondents of soybean commodity farmers in Jember regency in 2024.*

aspect, the average farmer in the Jember Regency still has a positive perception of soybean farming. This is proven by the fact that there are still farmers in the area who continue to plant soybean commodities even though some other farmers have switched their farming to other commodities. This is in line with the results of research by Rani et al. [28] in Simpang Village, Berbak District, East Tanjung Jabung Regency, which revealed that most farmer respondents perceived that soybean farming was considered easy to do.

Meanwhile, from the total average score of farmer responses as a whole (63.72), some include the aspect of farmer motivation with an average score of 10.46 with a score range between 3 and 15. This condition indicates that the motivation of farmers to plant soybean commodities on their land is still quite strong. Various factors can influence farmers' motivation in planting soybeans, including factors of experience, education, family responsibilities, land area and agroecosystem simultaneously influence the motivation of soybean farmers. While partially, the ones that have a significant influence are education, experience and family responsibilities, while the variables of land area and agroecosystem do not affect the motivation of soybean farmers [29]. The motivational aspect is a problem related to socio-psychology and is considered an important factor that influences human behaviour and performance. Motivation can influence farmers' adoption decisions and reflect the goals and objectives that farmers want to achieve in the long term, so that this motivational aspect will able to become better-performing farmers (Pannell et al., 2006, Farmar-Bowers and Lane, 2009 in Nurain et al. [29]).

Likewise, in terms of farmer knowledge of several important factors related to soybean commodities, the average score is 13.06 from a score interval of 5–25. The condition in this aspect is very severe; farmers do not have much knowledge about soybean farming, especially regarding government regulations, the amount of demand and supply of soybean commodities, developments in output prices and others. This aspect is very important in relation to the attitudes and skills of farmers in cultivating agricultural commodities. The low level of farmer knowledge about soybean cultivation causes the response to soybean cultivation to be quite low. According to Yuswandi et al. [30] that quality farmer knowledge is an absolute prerequisite for the success of agricultural development. Knowledge is the initial stage of perception, which then influences attitudes and in turn gives birth to actions or actions (skills). With good farmer insight into something, it will encourage attitudes, which in turn encourage changes in behaviour. Based on the results of his research in Pamboang District, Majene Regency, he revealed that the aspect of farmer knowledge has a positive effect on increasing farmer production, including soybean commodity production. This condition can be interpreted that if the knowledge of farmers in this research area becomes broader, especially regarding soybean cultivation and its economic aspects, then the response will be directly proportional.

**Table 4** also illustrates that the response of farmers, especially in the aspect of attitude towards soybean farming in Jember Regency, is known that the average number of scores is 11.18 from the interval score of 3 to 15. This means that based on

| No. | Socio-economic factors | Number of respondents (people) | Percentage (%) |
|---|---|---|---|
| 1 | Farmer perception: | | |
| | a. Soybean farming is economically less profitable | 27 | 54 |
| | b. Technically, soya farming is not easy to do | 3 | 6 |
| | c. Climatologically, there is no guarantee of a definite season | 9 | 18 |
| | d. By regulation, the government does not protect soybean farmers | 11 | 22 |
| | Total | 50 | 100 |
| 2 | Farmer motivation: | | |
| | a. Follows other farmers who are not interested in growing soybeans | 8 | 16 |
| | b. Still traumatised by past experiences | 17 | 34 |
| | c. Lack of encouragement from the government to engage in soya farming | 5 | 10 |
| | d. No policy and protection from the government | 9 | 18 |
| | e. No *floor* price for soya production | 11 | 22 |
| | Total | 50 | 100 |
| 3 | Farmer knowledge: | | |
| | a. Lack of knowledge about the risks of soybean farming | 7 | 14 |
| | b. Unknown amount of market demand for soya commodities | 15 | 30 |
| | c. Lack of knowledge of the development of soybean commodity prices | 18 | 36 |
| | d. Not aware of the government's policy on soya commodities | 10 | 20 |
| | Total | 50 | 100 |
| 4 | Farmer attitudes: | | |
| | a. Interested in staying in soybean farming | 10 | 20 |
| | b. Quite interested in soya farming | 15 | 30 |
| | c. Not interested in soya farming | 25 | 50 |
| | Total | 50 | 100 |

*Source: Primary data processed year 2024.*

**Table 5.**
*Socio-economic factors affecting farmers' response to soybean farming in Jember regency in 2024.*

the other three aspects, the attitude of farmers firmly states that they will continue to plant soybean commodities and also firmly states to leave soybean farming. Although the phenomenon leads to a weaker response of farmers to plant soybean commodities. This phenomenon also occurs in the Southern part of Banyuwangi Regency where no less than 50% of soybean farmers switch to other commodities such as citrus fruits, dragon fruits and other types of horticultural crops. The other part still persists to plant soybean commodities in the season until there are farmers who have to do land rental for soybean farming both on rice fields, yards and moorlands [6]. In contrast to the research results of Noormansyah and Dini [11] that the response of farmers in Ciamis Regency is classified as high at various scales of business. Regarding the socio-economic factors that influence the weaker response of farmers to cultivate soybean commodities can be seen in **Table 5**.

**Table 5** explains that most (54%) farmer respondents have the perception that soybean farming is economically less profitable than other commodities. The motivation of farmer respondents who are not interested in planting soybeans is mostly (34%) due to the fact that they are still traumatised by past conditions and some others are due to following other farmers who are no longer interested in cultivating their land planted with soybeans. On the other hand, despite the lack of encouragement from the government to cultivate soybeans, no policies and protection from the government and no *floor* price for soybean production, 50% of respondents continue to plant soybean commodities. This is because they still have hope that soybean commodities are still the main food ingredient as a source of protein for humans where the amount of consumption is still more than the existing stock.

**Table 5** also reveals that farmers' level of knowledge about the amount of market demand for soybean commodities, government policies on soybean commodities and the development of soybean commodity prices is relatively low. Lack of government attention is the main cause of the farmers' low knowledge. As a result of the above phenomenon, as many as 50% of farmer respondents took a firm stance not to be interested in cultivating soybean commodities again. This condition is in line with Haryono's opinion in Tempo that there are three factors hindering the increase in soybean production in the past year. First, there is no market guarantee for the price of local farmers' soybeans. Farmers are also unsettled by the invasion of imported soybeans that can be cheaper than local soybeans. This has caused farmers' lack of interest in planting soybeans. Secondly, the socialisation and distribution of superior seeds produced by the Research and Development Agency have not gone well. Farmers have not had enough access to use these varieties in their production fields. Third, there is a lack of government protection for local soybean farmers. This is indicated by the implementation of a zero per cent tariff policy on soybean imports. Soybean farmers have suffered losses because the price of their soybean production is lower than the price of imported soybeans.

Indeed, the potential of soybean farming in Jember Regency is very prospective, and this is in line with the results of research by Firdaus which revealed that soybean commodities in Jember Regency have a competitive advantage because the value of the *Private Cost Ratio* (PCR) is less than one, which is 0.8733, which means that to produce one unit of value-added output at private prices requires the sacrifice of domestic resource factors of 0.8733 units or to produce one unit of output can be saved by 0.1267 units. The prospective economic potential of soybean farming is not excessive because the results of research by Soetriono et al. [31] using the calculation of *Location Quotient (LQ)* analysis concluded that one of the bases of soybean crops in East Java Province is in Jember Regency. The results of Soetriono et al. [21] are in accordance with the results of the economic study of soybean farming conducted by Feriady which concluded that soybean farming is an attractive business opportunity to do both in terms of technical cultivation that is relatively easy to do, as well as market opportunities that are still very broad and economically still profitable financially very feasible to do.

## 3. Conclusions

1. The average age of farmers in Jember District is 49.26 years (classified as productive age), the average level of formal education of farmers is classified as moderate (9.42 years), the average number of household members of farmers in Jember District is 3.26 people (classified as a medium family), average land area controlled by farmers and planted with soybean commodities is 0.55 hectares (classified as quite large) and the average area of rice fields controlled by farmers as a whole in Jember District is 0.36 hectares (classified as narrow land area).

2. The average profit level of soybean farming is IDR 4,487,125.76 per hectare compared to the range of farm profit levels in Jember Regency which is low and the average R/C Ratio level of soybean farming reaches 1.83 and when compared to the results of previous research is also low but the tendency has the potential to increase.

3. The average response of farmers to soybean farming in the study area is classified as less strong or weak (score 63.72).

4. Socio-economic factors causing the weak response of farmers to plant soybean commodities are the factor that soybean farming is less profitable than other commodities and the absence of product price protection from the government.

## Acknowledgements

1. IntechOpen Publisher – London who has agreed to publish this research manuscript.

2. DP2M DIKTI Ministry of Research, Technology and Higher Education of the Republic of Indonesia for supporting the cost of this research, so that this research process can be carried out properly and smoothly.

3. Rector of Muhammadiyah University of Jember, Dean of the Faculty of Agriculture, Head of the Agribusiness Study Program, Colleagues and Students who have supported the process of implementing the research and compiling the book chapter manuscript.

4. Farmers and managers of farmer groups who have agreed to be respondents in this research.

5. Family Members of the Author Team who have supported the efforts to compile this book chapter manuscript.

## Author details

Syamsul Hadi*, Insan Wijaya and Anisa Nurina Aulia
Teaching Staff of Faculty of Agriculture, University of Muhammadiyah Jember, Indonesia

*Address all correspondence to: syamsul.hadi@unmuhjember.ac.id

## References

[1] Pusdasi Sekjen Kementan. Analysis of Soybean Trade Performance in 2023. Jakarta: Centre for Agricultural Data and Information Systems - Ministry of Agriculture; 2023. Available from: http://epublikasi.setjen.pertanlan.go.ld https://www.bps.go.id/id/publication/2023/11/22/240a70a50154f74d4e2381d0/distribusi-perdagangan-komoditas-kedelai-indonesia-2023.html

[2] Hulu A. Studi Inovasi Strategi Kebijakan Percepatan Pencapaian Swasembada Kedelai Indonesia Tahun 2035. Matra Pembaruan. 2023;**7**(1):13. DOI: 10.21787/mp.7(1).2023

[3] Santosa DA. Uncovering the Causes of Soy Scarcity in Indonesia. Webinar on Crisis Phenomena and Critical Points of Food Provision Organised by the Board of Professors of IPB University on 20 March 2022; 2022. Available from: https://ppid.ipb.ac.id/prof-dwi-andreas-santosa-ungkap-penyebab-kelangkaan-kedelai-di-indonesia/

[4] Satriani R, Rifda N, Akhmad RK, Eni HW, Alfina H, Dan Rifki AN. Farmer satisfaction level of institutional performance of soybean farming in Central Java. Jurnal Ilmu Pertanian Indonesia (JIPI). 2023;**28**(4):572-580. DOI: 10.18343/jipi.28.4.572

[5] Bardono S. Balitbangtan Develops High Productivity Soybeans in Jember. 2021. Available from: https://technologyindonesia.id/pertanian-danpangan/inovasi-pertanian; https://technologyindonesia.id/pertanian-dan-pangan/inovasi-pertanian/balitbangtan-kembangkan-kedelai-produktivitas-tinggi-di-jember/

[6] Hazmi M, Sutiarso E, Syamsul H. Efforts to fulfil stock of soybean commodity (Glycine max) through restoration of rural agribusiness development programme. International Food Research Journal. 2018;**25**(Suppl. 2):S127-S134. Available from: http://www.ifrj.upm.edu.my/25%20(08)%202018%20supplementary%202/2%20-%20IFRJ18147.R1%20edited.pdf

[7] Jember Regency Food Crops, Horticulture, and Plantation Service. DPTHP Jember Regency: Farmer Group Data, Number of Members, Leading Commodities, Administration & Agricultural Land Area in 2024 (Internal Document). Jember Regency: DTPHP Jember Regency; 2024. Available from: https://dtphp.jemberkab.go.id/

[8] Candra N, Dina A, Dan Zulfikar N. Faktor-Faktor yang Berpengaruh terhadap Motivasi Petani Kedelai di Kabupaten Tasikmalaya. Suluh Pembangunan: Journal of Extension and Development. 2021;**3**(1):53-60. DOI: 10.23960/jsp.vol3.no1.2021.125

[9] Zakaria AK. Soybean cultivation development policy towards selfsufficiency through farmer participation. Journal of Agricultural Policy Analysis. 2016;**8**(3):259-272. DOI: 10.21082/akp.v8n3.2010.259-272

[10] Widowati EH, Alfina H. Impact of government programmes on increasing soybean production in Central Java. Journal of Research and Development of Central Java Province. 2022;**20**(2):205-213. DOI: 10.36762/jurnaljateng.v20i2.1009

[11] Noormansyah Z, Dini R. Relationship between characteristics and farmers' response in soybean development programme (case in SL-PTT soybean programme in Ciamis regency).

Journal of Mimbar Agribusiness. 2016;**1**(2):149158. DOI: 10.25157/ma.v1i2.56

[12] BPS-Statistics Indonesia Jombang Regency. Rice, Corn, and Soybean Production. 2016. Available from: https://jombangkab.bps.go.id/en/pressrelease/2016/07/01/318/produksipadi-dan-palawija--angka-tetaptahun-2015-.html. 19Soybean Science and Sustainability – Advancing Crop Productivity, Nutrition, and Climate Resilience

[13] BPS Kabupaten Jember. Jember in Figures 2013. Jember: BPS Kabupaten Jember; 2014. Available from: https://search.yahoo.com/search

[14] Nazir. Research Methods. Jakarta: Ghalia Indonesia; 1985. Available from: https://books.google.co.id/books/about/Metode_penelitian.html?id=M_-dAQAACAAJ&redir_esc=y

[15] Singarimbun M, Sofian E. Survey Research Methods. Jakarta: LP3ES; 2008. Available from: https://www.scribd.com/document/394238385/12-Metode-Penelitian-Survei dpk.kepriprov.go.id › opac

[16] Koentjaraningrat. Methods of Community Research. Jakarta: PT Gramedia; 2008. Available from: https://www.scribd.com/document/861057071/Metode-Metode-Penelitian-Masyarakat-Koentjaraningrat; https://gpu.id/book/75927/metode-metode-penelitian-masyarakat; https://books.google.co.id/books/about/Metode_metode_penelitian_masyarakat.html?id=-xYGzQEACAAJ&redir_esc=y; https://openlibrary.telkomuniversity.ac.id/pustaka/3987/metode-metode-penelitian-masyarakat.html

[17] Soekartawi. Economic Theory of Production with the Subject Matter of Cobb-Douglas Function Analysis. Jakarta: CV Rajawali; 2002. Available from: https://lib.unj.ac.id/slims2/index.php?p=show_detail&id=8573; https://elibrary.univamedan.ac.id/index.php?p=show_detail&id=4201; https://www.kikp-pertanian.id/polbangtanyoma/opac/detail-opac?id=24763

[18] Meryani. Analysis of soybean farming and tataniaga in Ciranjang district, Cianjur Regency, West Java agribusiness management programme [thesis]. Bogor Agricultural University; 2008 (not published)

[19] Sugiana I, Sadguna D, Tonga Y, Kaca I. The study on farmers welfare. International Journal of Life Sciences (IJLS). 2018;**2**(1):29-41. DOI: 10.29332/ijls.v2n1.92

[20] Soekartawi. Prinsip Dasar Ekonomi Pertanian: Teori Dan Aplikasinya. Jakarta: Rajawali Press; 1998. Available from: https://unb.ac.id/perpustakaan/index.php?p=show_detail&id=1961; https://digilib.unikama.ac.id/index.php?p=show_detail&id=4980; https://digilib.ukwk.ac.id/index.php?p=show_detail&id=1640; https://www.kikp-pertanian.id/smkppsembawa/opac/detail-opac?id=529; https://onesearch.id/Record/IOS3558.slims-16802/Details

[21] Soetriono DS, Rita H, Ariq DM, Dimas BZ. Soybean Commodities, Agroindustry Centres and Technology Development Models in East Java. Malang: Intimedia; 2021. Available from: https://bintangpusnas.perpusnas.go.id/konten/BK51736/komoditas-kedelai-sentra-agroindustri-dan-model-pengembangan-teknologi-di-jawa-timur; https://bacabuku.com/detail/komoditas-kedelai-sentra-agroindustri-dan-model-pengembangan-teknologi-di-jawa-timur/51736; https://e-library.poliwangi.ac.id/index.

php?p=show_detail&id=20877; https://shopee.co.id/Komoditas-Kedelai-(Sentra-Agroindustri-dan-Model-Pengembangan-Teknologi-di-Jawa-Timur)-Soetrion o-i.121103960.14238935301

[22] Mantra IB. Pengantar Studi Demografi. Yogyakarta: Nurcahya; 1985. Available from: https://books.google.co.id/books/about/Pengantar_studi_demografi.html?id=AWIyAAAAMAAJ&redir_esc=y https://openlibrary.org/books/OL2703268M/Pengantar_studi_demografi

[23] BPS Kabupaten Jember. Jember in Figures Year 2024. Jember: BPS Kabupaten Jember; 2024. Available from: https://jemberkab.bps.go.id/id/publication/2024/02/28/06ff1b2512ce35 5752be8d2f/kabupaten-jember-dalam-angka-2024.html

[24] Indonesia R. Undang-Undang Republik Indonesia Nomor 24 Tahun 2013 tentang Perubahan atas UndangUndang Nomor 23 Tahun 2006 tentang Administrasi Kependudukan. Jakarta: Lembaran Negara/DPR; 2013. Available from: https://peraturan.go.id/id/uu-no-24-tahun-2013 https://peraturan.bpk.go.id/Details/38985/uu-no-24-tahun-2013

[25] Widotomo H, Arifin MZ. Efforts to increase soybean (Glycine max Merr.) production to increase farmer profits in East Java. In: J-SEP. Vol. 2, No. 1. 2008. Available from: https://www.e-jurnal.com/2014/06/upaya-peningkatan-produksi-kedelai.html and Kode Jurnal: jppertaniandd080044

[26] Zakaria AK, Sejati WK, Kustiari R. Analysis of soybean commodity competitiveness according to agro-ecosystems: A case study in three provinces in Indonesia. Jurnal Agro Ekonomi. 2010;**28**(1):21-37. DOI: 10.21082/jae.v28n1.2010.21-37 and License: CC BY-NC-SA 4.0

[27] Meryani. Analysis of soybean farming and tataniaga in Ciranjang District, Cianjur Regency, West Java [thesis]. Agribusiness Management Programme, Bogor Agricultural University; 2008 (not published)

[28] Rani, Siata R, Sardi dan I. Persepsi Petani Terhadap Usahatani Kedelai di Kelurahan Simpang Kecamatan Berbak, Kabupaten Tanjung Jabung Timur. Sosio Ekonomika Bisnis. 2012;**15**(2):9-18. DOI: 10.22437/jiseb.v15i2.2750

[29] Nuraini C, Agustina D, Noormansyah Z. Faktor-faktor yang berpengaruh terhadap motivasi petani kedelai di Kabupaten Tasikmalaya. Suluh Pembangunan: Journal of Extension and Development. 2022;**3**(1):53-60. DOI: 10.23960/jsp.vol3.no1.2021.125

[30] Yuswandi SS, Dan Andi DJ. Pengaruh Pengetahuan, Keterampilan dan Perilaku Terhadap Peningkatan Produksi Pertanian di Kecamatan Pamboang. SJM: Sparkling Journal of Management. 20Farmers' Responses to Soybean (Glycine max (L.) Merrill) Farming and the Influencing Socio.... 2023;**1**(3):255-267. Available from: https://e-jurnal.nobel.ac.id/index.php/

[31] Soetriono. The Policy Analysis Matrix (PAM). Jember: University of Jember Postgraduate Programme, Agribusiness Management Study Programme; 2002